Introducing Geology

Companion titles:

Introducing Palaeontology – A Guide to Ancient Life (2010)

Introducing Volcanology ~ A Guide to Hot Rocks (2011)

Introducing Geomorphology – A Guide to the Landforms and Processes (2012)

Introducing Meteorology ~ A Guide to the Weather (forthcoming 2012)

Introducing Tectonics, Rock Structures and Mountain Belts (forthcoming 2012)

Introducing Oceanography (forthcoming 2012)

For further details of these and other Dunedin
Earth and Environmental Sciences titles see
www.dunedinacademicpress.co.uk

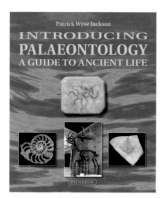

ISBN 978-1-906716-15-8

ISBN 978-1-906716-22-6

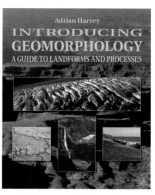

ISBN 978-1-906716-32-5

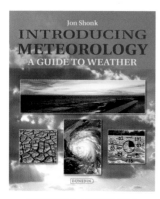

ISBN 978-1-780460-02-4

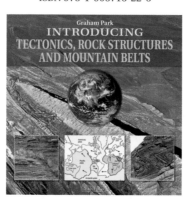

ISBN 978-1-906716-26-4

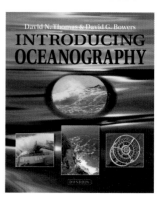

ISBN 978-1-780460-01-7

Introducing Geology

A Guide to the World of Rocks

SECOND EDITION

Graham Park

DUNEDIN

Published by
Dunedin Academic press Ltd
Hudson House
8 Albany Street
Edinburgh EH1 3QB
Scotland

ISBN 978-1-906716-21-9
© 2010 Graham Park

First published 2006, reprinted 2008, 2009
Second Edition 2010, reprinted 2012

British Library Cataloguing in Publication Data
A catalogue record for this book is available from the British Library

Design, pre-press production and typesetting
by Makar Publishing Production, Edinburgh
Printed and bound in Poland by Hussar Books

Contents

Acknowledgements

I am indebted to Mrs Anne Shelley of the Orcadian Stone Company, Golspie, Sutherland, for allowing me to photograph some of the excellent specimens of fossils and minerals in her geological museum. My thanks are also due to Anne Morton of Dunedin Academic Press and an anonymous reviewer of the draft of the first edition for their numerous helpful suggestions, and to Professor Charles Holland of Trinity College, Dublin for his careful review of chapters 9 and 11.

The second edition has benefited from a number of helpful suggestions for improvement from various reviewers of the first edition. I am grateful in particular to John Winchester, who pointed out a number of mistakes and drew my attention to the revised geological timescale. Any remaining inadequacies in this book are entirely the author's responsibility.

Finally I wish to thank my wife Sylvia for her constant support and encouragement, and as a non-geologist, for 'test driving' the first draft.

The following sources of data were particularly useful.

Duff, P. McL. D., *Holmes' Principles of Physical Geology*, 4th edition, Chapman & Hall, London, 1993.

Gradstein, F. M., Ogg, J. G. & Smith, A. G., *A Geologic Time Scale*, Cambridge University Press, 2005.

Keary, P. (ed.), *The Encyclopedia of the Solid Earth Sciences*, Blackwell, Oxford, 1993.

Lambert, D., *The Cambridge Field Guide to Prehistoric Life*, Cambridge University Press, 1985.

Stanley, S. M., *Exploring Earth and Life through Time*, Freeman, New York, 1993.

Note to the second edition

I have taken the opportunity to thoroughly revise the text and improve many of the line drawings, with greater use of colour, taking into account the many helpful suggestions made by reviewers. Some of the photographs have been exchanged for better versions, and a number of others added. The geological timescale has been updated according to a revised version published in 2005. *Graham Park, February 2010.*

List of tables and illustrations

List of tables and illustrations

Figure 1.2 A, **quartz** crystal, note hexagonal form; B, **calcite** crystal; C, **geode** containing purple amethyst; D, **muscovite mica**, note sheet-like crystal structure. B and C, courtesy of the Orcadian Stone Company Geological Museum, Golspie, Sutherland; D, British Geological Survey. ©NERC all rights reserved. IPR/122-06C.

1

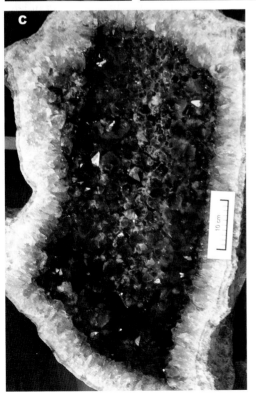

adorning many of our banks – a good place, incidentally, to look for examples of minerals. Another common mineral, also to be found in many granites, is **mica**, of which there are two main varieties, **biotite** and **muscovite**. Biotite is brown in colour, whereas muscovite is almost colourless (Figure 1.2D). Micas occur as thin sheets or flakes and their flat surfaces gleam in the light so that they are easily distinguishable from the accompanying feldspar and quartz. Muscovite is a hydrated potassium aluminium silicate, with the complex chemical formula $KAl_2(OH)_2(AlSi_3O_{10})$. Biotite is similar but contains iron and magnesium in place of some of the aluminium. The presence of iron is responsible for the darker colour of this mineral. The sheet-like or tabular form of mica crystals is due to their atomic structure (e.g. *see* Figure 1.1C), in which relatively strong silicate sheets are held together by weaker bonds consisting of potassium and aluminium atoms.

Granites are light coloured, usually pale grey or pink, and always contain quartz. The dark-coloured igneous rocks, however, are characterised (typically) by the absence of quartz and by the presence of one or more of three dark-coloured minerals in addition (usually) to feldspar; these are **olivine, hornblende** and **pyroxene**. Olivine is an iron-magnesium silicate with a relatively simple structure and has a glassy green appearance. Pyroxene has a similar chemical composition to olivine but may contain calcium and aluminium in addition. It ranges in colour from brown or dark green to black. Hornblende is the most complex of the dark rock-forming silicates, and also one of the commonest. It is a hydrated aluminium silicate containing variable amounts of calcium, magnesium and iron; its colour is usually black, sometimes with a greenish tinge. Hornblende and pyroxene are difficult to distinguish in rock samples and rarely show good crystal structure; however hornblende is usually more shiny in appearance and is more intensely black than pyroxene.

Calcite (calcium carbonate) (Figure 1.2B) is the main constituent of limestones and is common in other sedimentary rocks. Well-formed crystals of calcite are to be found in veins, often accompanying other less common ore minerals. **Halite** or **rock salt** (sodium chloride) is found in sedimentary deposits formed by the evaporation of sea water. Such deposits, termed **evaporites**, also include **gypsum** (hydrated calcium sulphate) (Figure 1.3A). **Barite** (barium sulphate) is a common vein mineral, often associated with calcite or quartz. Because of its high density, it is used by the oil industry to provide weight in drilling muds. Another common vein mineral is **fluorite** (calcium fluoride) (Figure 1.3B) which forms attractive purple crystals and occurs particularly in limestone areas.

Graphite (carbon) and **sulphur** are two examples of what are termed 'native elements', which occur in this form as well as in combination, as in oxides or silicates. Graphite requires high temperatures for its formation and only occurs in metamorphic rocks. Sulphur is usually bright yellow in colour and forms around active volcanoes and hot springs.

Ore minerals

An **ore** is a metal-bearing mineral, which may be commercially valuable if present in high enough concentrations. Ore minerals occur in small quantities within many igneous and sedimentary rocks but commercially useful amounts usually occur in bodies of rock such

Figure 1.3 A, **gypsum**; B, **fluorspar** (fluorite).
Courtesy of the Orcadian Stone Company
Geological Museum, Golspie, Sutherland.

as veins, where the ore is concentrated. Such mineral veins contain variable quantities of ore minerals along with much larger quantities of common minerals such as quartz and calcite from which the ore is extracted.

Iron ore minerals are the commonest and include iron oxide (**hematite** and **magnetite**) and iron sulphide (**pyrite**). Hematite (or haematite) (Figure 1.4A) is the most familiar ore with its characteristic reddish-brown colour, known to everyone in the form of rust. Magnetite is black and, as its name suggests, is magnetic. This property can easily be demonstrated by holding a magnet or compass needle close to the mineral. Pyrite (Figure 1.4B) is distinguished by its cubic form and brassy yellow colour ('fool's gold'!).

Copper ores include **chalcopyrite** (copper-iron sulphide) and **malachite** (copper carbonate). Chalcopyrite is very similar to pyrite but has a more 'coppery' colour. Malachite (Figure 1.4C) has a very characteristic green colour which is often seen on weathered copper roofs. **Azurite** (Figure 1.4E) is a hydrated copper carbonate, similar in composition to malachite but with a crystalline form and brilliant blue colour.

Other common ores are **galena** (lead sulphide) (Figure 1.4D) and **sphalerite** (zinc sulphide), both of which are often found in calcite veins within limestone. Ores consisting of metal only include native **copper** and **gold**. These are much less common; although gold is present in tiny quantities in many rocks, it can only be exploited commercially when it becomes concentrated either by sedimentary processes, in river gravels for example, or in mineral veins along with quartz (*see* Chapter 11). Even in veins, it is usually present in minute quantities and has to be extracted by crushing large quantities of rock and separating out the gold.

Gemstones

Gems, or **gemstones**, are minerals which are considered to be of special value because of their beauty, rarity and (in most cases) resistance to wear. Their commercial value depends

1

Figure 1.4 A, **hematite**; B, **pyrite**; C, **malachite**, D, **galena**, E, **azurite**. Courtesy of the Orcadian Stone Company Geological Museum, Golspie, Sutherland.

to varying degrees on all three of these factors. Thus there are minerals that can be very beautiful in their crystalline form, such as quartz, but only certain uncommon and attractively coloured varieties of quartz, such as the purple **amethyst** (Figure 1.2C), would qualify as gemstones. Thus rarity is an essential quality determining value. Resistance to wear is also important. Gems must be resistant to both physical wear and chemical deterioration. All the most highly prized gemstones, such as diamond, are as hard, or harder than quartz, and thus cannot be scratched by dust particles etc.

The most valuable gem minerals include diamond, sapphire, ruby and emerald. **Ruby** and **sapphire** are varieties of **corundum** (aluminium oxide), the red and blue colours being due to minute impurities of iron, chromium and titanium. **Diamond** is a high-pressure form of carbon, produced at depths of over 100 km and brought to the surface in certain igneous rocks (*see* Chapter 11). **Emerald** is a variety of **beryl**, which is a beryllium aluminium silicate, whose green colour is produced by the inclusion of small amounts of chromium within the beryl crystal structure.

Other gem minerals include **topaz** (a hydrated aluminium silicate including some fluorine), **zircon** (zirconium silicate) and **tourmaline**. Tourmaline has a very complex chemical formula – essentially a hydrated iron-aluminium-silicate, but with significant additions of boron and sodium. The rarity, and thus value, of the latter two gemstones results from the fact that they include the relatively rare elements zirconium and boron.

The shapes of the gemstones used in ornaments, such as diamonds, are not natural crystal shapes but are obtained by the highly skilled process of cutting the gem in a particular way to enhance its reaction to light. The cutting may take advantage of natural planes of cleavage within the mineral that are determined by its atomic structure.

Semi-precious stones

Gems are divided in the popular view into **precious stones**, which include all those valuable types mentioned above, and **semi-precious stones**. The latter, as the name suggests, are less valuable than 'precious' gemstones but include several popular ornamental minerals. Perhaps the best known of these is **agate**, which is a form of coloured quartz that is so finely crystalline that the individual crystals cannot be seen with the naked eye. In agate, the colours are arranged in bands, often concentrically around the inside of a geode. When polished, the smooth surface with its highly individual arrangement of grey, blue, purple or

pink colours make agate an extremely attractive ornament (Figure 1.5A). Other semi-precious gemstones include garnet, jade, amber and jet, together with coloured varieties of quartz crystals such as the purple **amethyst** (Figure 1.2C), smoky grey **cairngorm** and **rose quartz**.

Garnet (Figure 1.5B) is an aluminium silicate with varying proportions of iron, magnesium, calcium and manganese. The colour may be black, brown, red, or green, depending on which of these elements predominates. Garnet is found principally in rocks altered

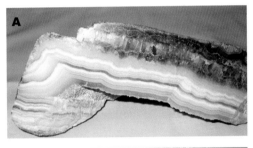

Figure 1.5 A, finely banded **agate**; B, **garnet** – dark-coloured, dodecahedral-shaped crystals in pale fine-grained schist. Courtesy of the Orcadian Stone Company Geological Museum, Golspie, Sutherland.

1

by heat and pressure (**metamorphic rocks**), formed at considerable depths, and usually forms quite large crystals with a good dodecahedral (12-sided) crystal form that is easy to identify. Its hardness and attractive colour make it most suitable for ornamentation – only its abundance reduces its value.

Jade is a form of pyroxene or **amphibole** in which sodium replaces the iron and magnesium in the commoner forms of these minerals. It has a beautiful green colour and is much prized by the Chinese for ornamental purposes because of its toughness.

Amber is solidified tree resin and is more properly regarded as a fossil substance than a mineral. However, it has a very appealing golden yellow colour and has been much used in making small decorative objects such as beads. It often contains fossil insects (Figure 9.5D) that have been trapped in the resin, which make it a valuable resource for palaeontologists. The relative softness of amber restricts its value as a gemstone.

Jet is another non-crystalline material that was formerly much used in the making of small ornaments and jewellery, especially by our prehistoric ancestors. It is organic in origin, being a form of coal, and is composed mainly of non-crystalline carbon particles. Black and shiny in appearance, its comparative softness makes it easily worked but renders it vulnerable to wear.

Volcanoes and melted rock

Volcanic eruptions can be both spectacular and highly destructive, warning us of the enormous power contained within the Earth and waiting to be released. Many will be familiar from recent television documentaries with the disastrous effects of explosive eruptions in the case of volcanoes such as Mount St Helens and Vesuvius. What may be less well known is the fact that much of the volcanic eruption that takes place around the globe does so quietly, through fissures, and that a large proportion occurs beneath the ocean surface.

Volcanoes and their effects

Volcanic eruption (Figure 2.1A) is the release of melted rock (**magma**) at the Earth's surface. The magma flows from a vent or fissure in the form of a slow-moving sticky liquid termed **lava** (Figure 2.1B). The speed at which lava flows can vary considerably since some lavas are more sticky (**viscous**) than others. The speed of even the fastest-flowing lavas down the slopes of a volcano is only a few kilometres per hour, and usually slow enough for people to escape and even for the stream of magma to be diverted into less damaging paths.

Highly viscous lavas are often accompanied by explosive eruptions, which emit particles of solid rock in addition to liquid magma. The particles vary from tiny (**volcanic ash**) to quite large blocks of broken rock (**volcanic breccia**) (Figure 2.2A). Pieces of **pumice** may be formed by the solidification of frothy magma rich in gas.

In highly explosive eruptions like that of Vesuvius in AD79, when the town of Pompeii was destroyed, a huge cloud of hot volcanic ash rises many kilometres into the atmosphere before descending and flowing at great speed down the slopes of the volcano, causing devastation to anything in its path (e.g. Figure 2.1C). In the case of the Vesuvius eruption described by Pliny the Younger, the descending cloud of ash and hot volcanic gases resulted in the burial of Pompeii in a deposit of ash up to 6 m thick and killing most of the inhabitants. It is thought that the bulk of the fatalities may have resulted from asphyxiation due to the hot gases, which generally include carbon dioxide and sulphur dioxide.

The more recent example of the eruption of Mount St Helens in the Cascade Mountains of north-western USA in 1980 has been studied in great detail. There, the eruption was preceded

2

Figure 2.1 A, an erupting volcano, Montserrat. IPR/73-34C British Geological Survey. © NERC. All rights reserved. B, ropy lava from a recent flow, Reykjanes, Iceland; C, pyroclastic flow entering the sea as a result of an explosive volcanic eruption, Montserrat. British Geological Survey. © NERC all rights reserved. IPR/122-06C.

by earthquake tremors two months before the eruption, followed by a bulging of the north side of the volcano to a height of over 150 m. After further vigorous earthquake activity, the bulge broke and collapsed, releasing a huge volume of gas and ash, and devastating a large area of the surrounding countryside. The immediate cause of the explosion in this case was thought to be the interaction of snow and wet rock with the hot magma. The force of the explosion was estimated to be about 500 times that of the Hiroshima atomic bomb.

Much more catastrophic eruptions have been recorded in the past, such as the one that destroyed the island of Krakatoa in 1883. There, a huge crater over 6 km across was formed on the site of the volcano and a vast glowing cloud of ash rose to a height of 80 km, gradually spreading around the globe in the upper atmosphere. It has been estimated that about 16 km^3 of rock was destroyed and that the ash fell over an area of about 4 million km^2. Even this event pales into insignificance compared to some of the events recorded in the geological past. There are examples in the geological record of eruptions covering areas of over 1000 km^2 with lava and around 20,000 km^2 with volcanic ash to a thickness of 20 m.

Flows of volcanic ash can reach speeds of tens of metres per second and are therefore much more dangerous than the slow-moving lava flows. Moreover the release of enormous quantities of ash and gases into the atmosphere can have significant adverse effects on the climate for months or even years after the eruption. Another hazard associated with explosive eruptions is the formation of huge waves called **tsunamis** that in the case of the Krakatoa eruption reached a height of 36 m and caused about 36,000 fatalities.

In some volcanic areas, activity is confined to hot springs and **geysers**, such as the famous 'Old Faithful' geyser in Yellowstone National Park in Wyoming and the type locality of Geysir in Iceland (Figure 2.2B). Here the source of heat is hot magma, which lies beneath the surface and heats the groundwater causing explosive eruptions of steam and other gases. Similar areas of hot springs have been used industrially as a source of heating in New Zealand and Iceland.

2

Figure 2.2 A: **volcanic breccia**; note coin for scale. B: **geyser,** from the type locality, Iceland.

Prediction of volcanic eruptions

All volcanic eruptions on land in recent times have occurred from existing volcanoes, many of which are known to be intermittently active. Those volcanoes that have not been active historically are known as **dormant** volcanoes, but even these cannot be regarded as completely safe from future activity. Active volcanoes near centres of population are generally closely monitored and some warning is usually possible of an impending eruption. Signs of increasing activity may continue for weeks or months before an eruption, and generally increase in intensity immediately beforehand, but the exact time and severity of the event cannot be forecast. Warning signs include swelling of the ground, minor earthquakes, explosions and the emission of clouds of gas. However, many eruptions, including some of the most damaging of recent times, have occurred from un-monitored sites and have taken the world completely by surprise.

Fissure eruptions

Although not as spectacular, nor dangerous, as the explosive eruptions just described, fissure eruptions are actually much more important in terms of the volume of lava produced over time. Fissure eruptions are the principal mode of formation of the oceanic crust and occur at present along the system of **ocean ridges**. In the Atlantic Ocean, for example, the mid-Atlantic ridge occupies about one-third of the surface area of the ocean, reaching a height of 2 km from the ocean floor. Only the central part of the ridge, however, is active and reaches the surface in the island of Iceland, where the volcanic processes may be more easily studied. Unlike volcanoes, where activity is centralised around a more or less circular vent, fissure eruptions take place along a crack that has been opened up by stretching (extension) of the Earth's crust. Fissures range from hundreds to thousands of metres in length and the lava pours out and flows laterally away, controlled by the shape of the ground surface. Later stages are usually marked by

Figure 2.3 A, igneous **dyke**; the host rocks are softer and have been eroded so that the dyke has the appearance of a wall (author for scale). B, basalt **pillow lava**; hammer for scale; note that the lower surfaces of the pillows are concave, resulting from their wrapping around the pillows beneath.

the formation of small dome-like structures (**shield volcanoes**) located at intervals along the fissure. When the magma solidifies within the fissure, it forms a structure known as a **dyke**, which is a sub-vertical sheet of igneous rock (Figure 2.3A). The whole area of Iceland is formed out of successive dyke intrusions and the lavas that have issued from them. Where the lava is poured out in water, it usually forms tube-like or pillow-like shapes that are termed **pillow lavas** (Figure 2.3B). These structures are formed by the rapid chilling of the lava forming a hardened skin around the outside of the structure within which still-molten lava may continue to flow.

Fissure eruptions also occur on the continents, although examples of these are known mainly from the geological record. In the Deccan region of India, for example, an area of $500,000\,km^2$ was covered by basalt lavas up to several kilometres thick in only a few million years during the Cretaceous Period. Fortunately, nothing on a comparable scale has occurred in historical times.

Channels and chambers

Magma, once formed, is lighter than the surrounding solid rock and is forced to migrate upwards under gravitational pressure. As it does so, small quantities of magma will collect together to form larger bodies and some will ultimately reach the surface in the form of lava. The precise way in which this migration takes place is dependent on a number of factors, including the composition of the magma, the nature of the rock being invaded, and the structural environment of these rocks, for example whether they are being stretched or compressed. If the magma is under high pressure, it may go straight up to the surface; if the magma pressure is lower, or if the rocks of the crust are more resistant to penetration, the magma may halt its upward journey part-way through the crust and form a **magma chamber**.

As mentioned above, a number of different types of structure can be formed by the intrusion of magma in this way. A narrow, steeply inclined, sheet-like structure is called a **dyke**. Good examples may be seen in many places in the coastal regions of western Scotland, especially on the islands of Arran, Mull and Skye, and on the Ardnamurchan peninsula. Since the dyke rock is often harder than the surrounding rock (**host rock**), which may therefore wear away more easily, the dyke may appear like a stone wall (Figure 2.3A). Large numbers of parallel dykes, forming a **dyke swarm**, are typical of regions of the crust undergoing stretching, or extension. Dykes may also be arranged radially around the base of a volcano.

A **sill** is a sheet-like structure, with a shallow inclination, intruded parallel to the planes of weakness in the surrounding rocks. The Great Whin Sill of northeast England, along part of which Hadrian's Wall is built, and the Stirling Castle Sill in central Scotland, are two well-known examples. In both these cases, the sill is much harder than the country rocks that it intrudes and stands out as a prominent escarpment. Both dykes and sills may be thought of as channels through which magma makes its way through the crust (Figure 2.4A). Some of the magma goes through the channel on its way upwards and some will eventually solidify in the channel to form a body of igneous rock (Figure 2.5C). Another type of channel is a **pipe**, or **vent**, which is a roughly cylindrical structure feeding a volcano.

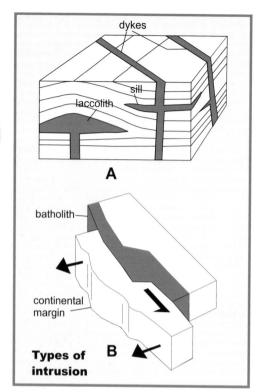

Types of intrusion

Figure 2.4 Magma chambers and channels – types of igneous intrusion. A. **dykes** and **sills** are formed by magma filling fissures formed by the stretching of the surrounding rock (host rock). Dykes are steeply inclined and cut across structures in the country rock; sills are gently inclined and intrude planes of weakness in the country rock; a **laccolith** is formed in a similar way to a sill but the surrounding rock is arched up to form a much larger structure. B, a **batholith** is a large magma chamber formed gradually over many thousands of years by successive intrusions as the host rock moves aside to make space for them.

These three types of structure are characterised by being narrow in comparison to their length. Dykes are typically only a few metres in width and rarely more than 100 m. Magma chambers, on the other hand, are much larger bodies, known as **plutons**. A **laccolith** (Figure 2.4A) is a sill-like body but is more lens-shaped than tabular and may reach a thickness of kilometres rather than metres. The largest plutons are known as **batholiths** (Figure 2.4B) and are found in the core regions of mountain belts such as the Cordilleran System of western North America. The British Columbia batholith, for example, is over 1000 km long and tens to hundreds of kilometres wide.

There has been much debate in the past about how large plutons make space for themselves within the crust; this was termed the **space problem**. It is now generally believed that most large magma bodies are intruded into, or along, crustal weaknesses and that the invaded rocks move aside to create the required space, in the same manner that dykes are injected into fissures that open gradually to make room for them.

Igneous rocks

There are a large number of different types of igneous rock, although many of them are uncommon or indeed rare. These differences are due to differences either in mineral content, or in mineral size, that is, in the grain size of the igneous rock.

Differences in mineral content result from variation in chemical composition, while differences in grain size are due to variation in the rate of cooling. Magma that cools quickly produces small crystals, and very rapid cooling produces glass, which is non-crystalline. Slowly-cooled magma, on the other hand, produces larger crystals, since the crystals have a longer time in which to grow before the magma solidifies. Consequently, large bodies of magma with a considerable store of internal heat take a long time to cool down and produce a coarse-grained rock, such as **granite** (Figure 2.5A). Large granite batholiths are known to take many thousands of years to solidify. Small bodies of intrusive igneous rock, such as dykes and sills, on the other hand, may cool and solidify comparatively rapidly, in a matter of years perhaps. Lavas, which are poured out on the Earth's surface, are subject to the most rapid cooling, and typically solidify within hours or days.

Examination of the margins of dykes and sills shows that there is usually a narrow zone of very fine-grained rock next to the cool country rock where the rate of cooling of the magma has been greatest. This zone is known as a **chilled margin**.

The simplest classification of igneous rocks depends on the recognition of two main

2

Figure 2.5 A, polished slab of **granite**, Aberdeenshire, Scotland, showing crystals of pink **feldspar**, dark **biotite mica** and glassy **quartz** (~8cm across); B, the cliff is formed from recent **basalt** at the mid-Atlantic ridge, Thingvalla, Iceland. C, dyke of fine-grained **dolerite** cutting banded **gneiss**, Orust, western Sweden. D, photomicrograph (× 2.5) taken in polarised light of an olivine gabbro, showing black and white **plagioclase feldspar** laths, pale grey **pyroxene** and brightly coloured **olivine**. A, D, British Geological Survey. ©NERC all rights reserved. IPR/122-06C.

sources of variation – grain size, and proportion of light to dark minerals. Thus rocks are divided into coarse-grained (average >5 mm), medium-grained (1–5mm) and fine-grained (<1mm), based on their grain size, and into **acid**, **intermediate** and **basic** based on the decreasing proportion of silica and the increase in dark (or ferro-magnesian) minerals. The table shows this simple classification. Broadly speaking, coarse-grained rocks are found in plutons, medium-grained in dykes and sills, and fine-grained in lavas, although both dykes and lavas may vary from fine to medium grained depending on their thickness.

In traditional geological usage, which is followed here, the terms 'acid' and 'basic' are used in a different sense from their standard chemical usage, to describe magmas respectively rich in (oversaturated with), and poor in (undersaturated with), silica. Some authors prefer the alternative terms salic (rich in Silica and ALumina) and mafic (rich in MAgnesia and iron (F)).

Acid rocks are characterised by an abundance of silica, in sufficient quantities to combine with all the other elements present to form silicates and leave enough over to form quartz. Of the dark (ferro-magnesian) minerals, only biotite or hornblende may be present and only in relatively small proportions. The other light mineral is feldspar, which makes up a large proportion of the rock. **Granite** (Figure 2.5A) is a coarse-grained acid rock and **rhyolite** a fine-grained type.

Intermediate rocks have lower proportions of silica, and free quartz is either absent, or present in very small quantities. The bulk of these rocks is formed of feldspar and a ferro-magnesian silicate, usually biotite or hornblende, or sometimes pyroxene. **Diorite** is a coarse-grained intermediate rock and **andesite** a fine-grained type.

Basic rocks have even lower proportions of silica and contain the ferro-magnesian minerals olivine and/or pyroxene, plus smaller amounts of feldspar. **Gabbro** (Figure 2.5D) is a coarse-grained basic rock, **dolerite** (Figure 2.5C) a medium-grained type and **basalt** (Figure 2.5B) the fine-grained type.

Ultrabasic rocks mainly contain the ferromagnesian silicates olivine and pyroxene, and feldspar is usually absent. **Peridotite** is a coarse-grained ultrabasic rock.

A wide variation in igneous rock types can be produced by the many permutations of the main rock-forming minerals already mentioned, or by the addition of other minerals introduced because of unusual amounts of a particular chemical element. The main silicate minerals themselves vary considerably in chemical composition. Thus feldspars range from potassium-rich varieties to sodium- or calcium-rich varieties, depending on the rock chemistry. Acid and intermediate rocks can be further subdivided according to whether potassium-rich or sodium–calcium-rich feldspars predominate. The potassium-rich equivalent of diorite, for example, is called **syenite**.

The Earth's interior and source of heat

In the late nineteenth century, it was generally believed that the Earth was only 20–80 million years old, based on a calculation by Lord Kelvin, who assumed that Earth had cooled to its present state from an initially hot, largely molten, body. However, the discovery of radioactivity led to the realisation that most of the heat currently being released through the Earth's surface must be derived from the breakdown of radioactive elements known to be present in significant quantities, principally uranium, potassium and thorium. Since heat is being continuously produced within the Earth, the present temperature distribution cannot be used to calculate its age. However, dating of rocks based on rates of radioactive decay showed that the oldest rocks were around 4000 million years old, and we now believe the age of the Earth to be 4600 million years old (*see* Chapter 8).

The interior of the Earth is divided into three regions (Figure 2.6). There is an outer thin shell called the **crust**, which varies in thickness from about 7 km beneath the oceans, to an average of about 35 km in continental areas, reaching nearly 80 km beneath certain young mountain belts. The innermost, approximately spherical, region is called the **core**, and extends from a depth of 2900 km to the centre of the Earth, at 6370 km depth. The region lying between the crust and the core is termed the **mantle**, and it is here that the processes that largely control what happens at the Earth's surface originate.

The composition of the crust is known in some detail, since material from all depths of the crust is directly accessible at the surface somewhere. There is a fundamental difference between oceanic and continental crust. Oceanic crust is composed almost entirely of the volcanic rock basalt, whereas continental rocks are extremely varied in composition, including the whole range of igneous rocks together with all the different types of rock derived from them. However, the average composition of continental crust has been estimated to be similar to a mixture of granite and basalt.

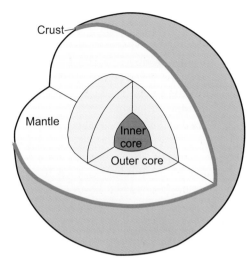

Figure 2.6 The cutaway model shows the main regions of the Earth's interior: solid inner core, liquid outer core, mantle and thin crust.

The core is believed to consist mostly of metallic iron, with some lighter elements in addition, such as nickel. In the outer core the metal is in a molten state but the inner core is solid. These constituents are thought to have become molten at an early stage in Earth's history and drained down towards the centre forming the core.

There is good evidence that the upper mantle is made of the ultrabasic rock peridotite (Table 2.1). This material is known to melt to form basalt magma, and pieces of peridotite are found within some basalts. Moreover, in places where oceanic crust has been thrust onto continental margins, mantle material consisting of peridotite is found directly beneath the crustal rocks.

Peridotite is composed mainly of the minerals olivine and pyroxene, which are silicates of iron and magnesium (*see* Chapter 1). Other minerals probably also present in mantle

grain size	acid	intermediate	basic	ultrabasic
coarse	GRANITE	DIORITE	GABBRO	PERIDOTITE
medium	MICRO-GRANITE	MICRO-DIORITE	DOLERITE	
fine	RHYOLITE	ANDESITE	BASALT	
main minerals	feldspar quartz mica hornblende	feldspar hornblende biotite	feldspar pyroxene olivine	pyroxene olivine

Table 2.1 Main types of igneous rock, distinguished by grain size: coarse (average > 5mm), medium (1–5mm), fine (<1mm); and by silica content: acid (characterised by excess silica, giving quartz; intermediate (characterised by abundant feldspar plus silica-rich ferromagnesian minerals hornblende and biotite); basic (feldspar plus silica-poor ferromagnesian minerals pyroxene and olivine) and ultrabasic (little or no feldspar plus silica-poor olivine and pyroxene).

peridotite include feldspar and metallic oxides. All the other elements found in the crust, including the heat-producing radioactive elements uranium, thorium and potassium, must also be present in small quantities, either within the main silicate minerals or in other compounds such as oxides, since the crust has been formed over time from magmas derived from the mantle.

The origin of magma

Melting of solid rock to produce magma takes place initially at depths of between 50 and 200 km beneath the surface. Once magma has been produced, it travels upwards and, because of its great heat, may cause further melting of rocks at higher levels. Such secondary melting, however, is comparatively insignificant in terms of volume.

In the popular imagination, hot magma is often conceived as welling up from some deep, permanently molten, layer within the Earth. Such a view is quite incorrect. As stated previously, the Earth's mantle is solid to a depth of around 2,900 km. Although below this level, the outer core consists of iron in a molten state, this material has no direct link with surface vulcanicity.

The temperature within the Earth increases downwards from the surface and has been estimated to reach about 1200°C at around 100 km depth beneath the continents, and at much shallower depths of less than 50 km in volcanically active oceanic regions. The rock at these depths is peridotite, and such a rock has been shown by laboratory experiments to begin melting at 1100°C at 50 km depth and at rather higher temperatures, around 1150°C, at 100 km depth (since melting temperature increases with increasing pressure). Peridotite melts only partially at temperatures of around 1200°C, since those components of the rock with the lowest melting temperature will melt first.

The layer of peridotite rock between 100 and 200 km depth is considered to contain small fractions (about 0.1%) of melt, which

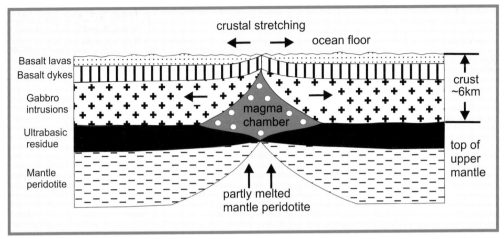

Figure 2.7 Origin of oceanic basaltic lavas. The diagram shows how the oceanic crust is formed by partly melted mantle **peridotite** moving up to form a magma chamber within the oceanic crust, aided by the stretching of the crust. The magma then separates into an **ultrabasic** residue, which sinks to the base of the chamber, and a **basalt** liquid, part of which solidifies within the crust as **gabbro** intrusions, and part of which moves up to the surface to form **basalt lavas** through narrow fissures that become dykes.

make the rock softer and more able to undergo slow flow in the solid state. These movements within the Earth's mantle, known as **convection currents**, are part of the process of plate tectonics, and will be explained in Chapter 5. Under certain conditions, if the Earth's crust is undergoing stretching for example, upward flow of partly molten peridotite may take place, leading to a reduction in pressure and a consequent increase in the proportion of melt. The liquid formed from a 1–15% melt of peridotite is **basalt**, and is the most abundant type of volcanic lava, making up almost all of the oceanic crust (Figure 2.7).

Causes of variation in igneous rocks

If magmas originate from the melting of mantle peridotite to form basalt, as has just been explained, why is it that so many different types of igneous rock occur? The answer to this question is to be found by examining two different processes that take place within the Earth: the first is termed **magma differentiation** and the second is the re-melting of previously formed igneous rock.

Magma differentiation is a process of physical separation that produces a new magma different in composition from the original one. Imagine a magma chamber filled with basalt magma undergoing slow cooling (Figure 2.7). The crystals with the highest melting temperature (olivine) will solidify first and, because they are heavier than the melt, will sink to the bottom of the chamber. The remaining magma will be depleted in the chemical components making up the olivine; it will be richer, for example, in silica. This process is called **fractional crystallisation**. This is the reverse of the process of **fractional melting** that is responsible for producing the basic rock basalt from the ultrabasic rock peridotite. If this differentiated magma is tapped and allowed to travel upwards and solidify separately, it will have a quite different composition to the original basalt, and may be intermediate or even acidic. Thus an igneous rock of granitic composition has been produced from a magma of basaltic composition, and a rock composed of olivine (peridotite) has been left in its place. However, only about 10% or less of the original magma can result in a granitic rock, and so this process cannot explain the abundance of granite in the Earth's crust and an alternative explanation must be sought.

Re-melting of solid igneous rock can take place in the presence of a suitable heat source.

Thus hot magma may melt and incorporate small amounts of solid rock with a lower melting temperature than the magma, so that basic magmas may melt acid rocks, but the reverse cannot take place. A basalt magma may therefore acquire a more intermediate composition by incorporating quantities of acid rock. The generation of large amounts of granitic and intermediate magmas, however, is usually considered to result from the physical depression (**subduction**) of basaltic rocks of the oceanic crust down into hotter regions of the mantle where they are subjected to fractional melting. This process takes place at subduction zones and plays an important role in **plate tectonics**, which will be described in Chapter 5.

Shaping the Land

There is a wonderful variety of the landscape, even within the British Isles, and the more so if we consider the differences between the polar regions, the North African deserts, and the great mountain ranges of the Alps and the Himalayas, for example. This great variation is the product both of climatic differences and of the interplay of the various forces that mould and modify the land surface. Understanding how these forces operate at the present day is important in interpreting rocks and geological phenomena of past ages. This important principle was first clearly stated by James Hutton in 1785 as 'the past history of our globe must be explained by what can be seen to be happening now', or in its more usual form 'the present is the key to the past'.

The land surface is a complex of **landforms** – mountains, valleys, plains, coasts, and so on, that have been produced by physical processes transporting rock material from one part of the land surface to another. For these processes to work, rock masses have to be uplifted above sea level, and so geological history is the record of continual creation of new land surfaces and their subsequent destruction. These destructive forces include the **weathering**, **erosion** (wearing down), and transport of rock material, and new rocks (**sedimentary rocks**) are formed by the deposition of this material elsewhere in the landscape or in the sea.

Weathering

Weathering is the process by which the constituent particles of a rock are changed in such a way that they are more easily broken down and worn away by erosion. The process of breakdown may be mechanical, chemical or biological in nature, or some mixture of these.

Mechanical weathering may be caused by the repeated alternations of expansion and contraction due to extremes of temperature, found for example in desert regions, which produce weakening of the rock along the grain boundaries and the eventual separation of the rock particles. **Frost shattering** is a similar process that works by the action of water expanding as it freezes within cracks in the rock. This process is one of the most important causes of erosion in mountain regions and can easily be observed after a frost where the surface of soil or gravel is raised up and the particles separated. Another type of mechanical weathering is often seen in areas

3

of granite rock which have been uplifted from deeper levels of the crust where the pressure is much greater. The reduction in pressure as the rock reaches the surface causes expansion to take place, opening up cracks parallel to the surface and resulting eventually in the rounded mounds known as **granite tors** (Figure 3.1).

Chemical weathering is caused by the chemical action of acidic rainwater on rock minerals. Most rainwater is slightly acid due to dissolved atmospheric carbon dioxide. Acid water may dissolve certain minerals, even quartz, but only in minute quantities. However, a more important effect is to alter the chemical composition of certain minerals, making them more susceptible to breakdown. For example, iron-rich silicates can be oxidised, producing a crust of red hematite on the surface and along cracks within the rock. Alumino-silicates such as feldspar break down to produce **clay minerals**; the reaction with water is known as **hydration**, which causes an expansion of the rock, and this is one of the most important methods of chemical weathering. Of the main

rock-forming minerals, quartz is the least susceptible to chemical weathering, followed by muscovite mica (in the form of small silvery flakes), which explains why these two minerals are so common in sand. Both mechanical and chemical weathering can be caused or aided by organisms; thus tree roots may exploit cracks to force quite large blocks of rock apart and lichens gradually dissolve the rock on which they grow.

Resistance to weathering varies according to the composition of the rock and the nature of the climatic conditions. Generally speaking, igneous and metamorphic rocks, which are formed of tightly interlocking crystals, are more resistant to weathering than sedimentary rocks composed of more loosely-bound, rounded, grains. Quartz sandstones are more resistant than fine-grained shales rich in clay minerals. Such variations in resistance are reflected in the landscape by the tendency of resistant rocks to form upstanding features, as shown, for example, by the prominent escarpments of the Grand Canyon (*see* Figure 4.1B).

Figure 3.1 Erosion: granite tor showing typical rounded outcrop with horizontal cracks (see text), Cairngorm mountain, Scotland. IPR/73–34C British Geological Survey. © NERC. All rights reserved.

Erosion

Erosion is the process whereby rock material that has been loosened and disintegrated by weathering is worn away by physical means and transported elsewhere. The disintegrated pieces of rock range in size from enormous boulders through gravel-sized fragments to sand-sized and clay-sized microscopic particles. These rock fragments are removed and transported by the action of water, wind and ice. An excursion into any mountainous area will reveal abundant evidence of the process of erosion. During heavy rain, sand-sized particles are washed downhill along channels worn by the flowing water; good examples of channelling can be seen on the beach as the tide recedes (Figure 3.2A). Strong winds transport fine dust or even sand; and in higher mountains with active glaciers, the ability of ice to transport quite large boulders is clearly evident. Larger rock fragments tend to be concentrated at the bottom of steep slopes forming aprons of debris, or **scree**, and **debris fans** typically are found at the bottom of steep gullies (Figure 3.2B). This material is initially transported mainly by gravity and further movement requires the action of a river or glacier system.

Figure 3.2 Erosion: A, effects of erosion by flowing water on soft beach sand; B, debris apron (**scree**) at the foot of steep basalt cliffs, Cuillin ridge, Skye, NW Scotland.

River systems

A river from its outlet in a sea or lake to its source, and including all its tributaries, constitutes a **river system**. The flow of water from source to outlet is governed by the difference in height between these two points and is known as the **head of water**. The gradient or slope of the ground at any point along the system controls the rate of flow and is important in determining how active the river is in transporting rock particles. Fast-moving rivers near their mountain source will move even large boulders while in flood, and in doing so will gradually wear them down into more rounded shapes, eventually forming well-rounded pebbles. In places where the slope eases and the river flow decreases, material is deposited as sediment, because the water is no longer able to carry it.

The wearing down of rock fragments, or **clasts**, is known as **abrasion**, and is carried out not by water alone but mainly by the rubbing action of one piece of rock on another. This effect can also be observed in the smooth rock surfaces of the river channel in fast-flowing sections. **Potholes** may often be seen where pebbles have been swirled round inside a depression, gradually smoothing and polishing the sides of the depression to form a bowl, and testifying to the abrasive action of the rock fragments under the influence of fast-flowing water (Figure 3.3A). Only the smallest particles of clay or mud are carried in suspension, while particles ranging from sand to boulder size are transported by sliding, rolling or bouncing along the river bottom. The smaller the size of the particles, the more easily they can be carried along by the flow of the river, which causes a separation of the clasts according to their size, the smallest being transported furthest. This process is called **sorting** and has the effect of dividing the river system into sections characterised by the deposition of a particular clast size; some parts by mud, some by sand and some by pebbles or boulders. This process is important in understanding how the different types of sedimentary rock are formed.

Rivers and streams normally flow within the confines of a channel that they have cut into the surrounding ground, although they may overflow their channel in times of flood. As more and more tributaries join the main river, as it progresses downstream away from its source, the volume of water increases, and so consequently does the width and depth of the channel. Narrow channels tend to promote turbulence in the water flow, which causes more active abrasion of the sides of the

Figure 3.3 River systems: A, pothole in fast-flowing mountain river, Braemar, Scotland. B, river meanders on the flood plain of upper Glen Spean, Inverness-shire, Scotland. IPR/73–34C British Geological Survey. © NERC. All rights reserved.

channel and facilitates the transport of larger rock fragments. As the river progresses downstream, the channel broadens, the gradient and rate of flow decrease, and there is a greater tendency for sediment to be laid down rather than carried along. In the lower stretches of a river, the channel typically forms a series of loops known as **meanders** (Figure 3.3B), which gradually migrate sideways because of the increased speed of the flow, and consequent increase in erosive power, on the outside of the curve. This sideways migration may reach a point where it becomes easier for the river to take a shortcut through the neck of the meander and the abandoned meander becomes what is known as an **ox-bow lake**.

When the river floods, the water spills over the side of the channel and spreads over the surrounding ground, leaving a layer of sediment. A wide area of relatively flat land subject to repeated flooding of this nature is termed a **flood plain**, and the lower reaches of major river systems are characterised by flood plains that may be many kilometres, or even hundreds of kilometres, wide. Over long periods of time, the course of such rivers may change many times to occupy different positions on the flood plain.

Youth to maturity in a river system

As time passes, more and more of the land surface is worn down by erosion, and although the process may take many thousands of years, eventually the land surface will become almost flat and close to sea level. At this stage the land surface is said to form a **peneplain** (almost a plain) and little further material is available to be transported to the sea by the river systems. This is the mature stage in the evolution of the landscape and its rivers. A 'young' landscape, in contrast, is marked by mountainous terrain and a large height difference between the land surface and sea level. A river in its youthful stages is fast flowing, its valleys are steep-sided, and large amounts of clastic debris are carried by it.

Most river systems at the present day are at some intermediate stage of maturity and exhibit a variation in characteristics. Near their source, the tributaries may show 'youthful' characteristics, set in steep-sided V-shaped valleys, whereas near their outlet to the sea they may occupy broad meandering channels set in wide valleys with extensive flood plains. A typical river profile at this stage will have steeper gradients near the source and gentle gradients near the outlet, as shown in the diagram (Figure 3.4). The more mature a river is, the smoother will be the profile and the gentler the gradient at source.

Uplift and rejuvenation

Rates of erosion, even in the most rapidly denuding mountain ranges, are only a few millimetres per year. A simple calculation shows that, at a rate of one millimetre per year, the highest mountain range will be completely levelled in only 10 million years, whereas the geological record shows that the oceans and seas of the Earth have been accumulating sediment for over 4000 million years! The explanation for this is that repeated crustal movements have uplifted the land surface, creating new mountain ranges and rejuvenating the river systems. **Rejuvenation** may take place either at the end of a cycle of erosion, when a young river system will be imposed upon an uplifted plain, or it may be

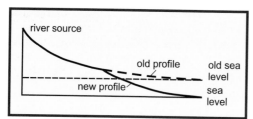

Figure 3.4 Effects of rejuvenation on a river profile caused by a fall in sea level.

3

superimposed on a mature system, in which case young V-shaped river valleys will cut into older, broader valleys. Where a rejuvenated river cuts through an older flood plain, flat **river terraces** are created above the level of the new river course.

A rejuvenated river profile will show a 'step' or sudden change in gradient where the new profile intersects the old (Figure 3.4). This step will migrate back towards the source with time until the new profile becomes completely established. Such rejuvenations are important in the sedimentary record because they are marked by a change in the nature of the sediment being deposited, from finer to coarser.

Coastal effects

The coasts bordering mature landscapes are characterised by low-lying plains and the river systems end in broad deltas where sediment carried down by the rivers is washed far out to sea. Marine erosion is usually limited to the action of currents operating parallel to the shoreline that carry sand laterally along the coast, removing it from one part and depositing it at another. This process is called **longshore drift** and its results can be seen in many parts of eastern and southern England in the form of long sand spits or bars extending across the entrances of estuaries or bays.

The coasts of rejuvenated landscapes on the other hand are characterised by cliffs, where coastal erosion may be quite marked, and indeed in historical times have resulted in the destruction of buildings. Cliffs are attacked by the action of waves that hurl pebbles and boulders at the base of the cliff and undercut them. Erosion also causes landslips from the top of the cliffs as the rocks become weakened by percolating rainwater from above and undercutting from below. Uplift of the land surface has the same effect as a lowering of sea level. In both cases, new land is created and may be demonstrated by the presence of **raised beach** levels. Many examples of raised shorelines at levels of around 5 and 8 metres respectively above present mean sea level are to be seen around the coasts of Scotland (Figure 3.5A). These were formed when the land surface rose as a result of the removal of the ice sheet at the end of the last Ice Age. Each of these beach levels varies in height, becoming higher when followed inland along estuaries and lower on the offshore islands. This is because the land surface became domed upwards, rising higher where the ice sheet had been thickest. Even higher raised beaches result from earlier (inter-glacial) periods of retreat of the ice sheet.

Close inspection of modern beaches is very useful in understanding geological processes, and many of the features seen in sedimentary rocks can be explained by studying them. One of the most important observations is the high degree of sorting by sediment size. Thus beaches tend to be either muddy or sandy or pebbly, and if pebbles and mud co-exist, they do so in alternate layers. This is because the action of the tides moves the smaller particles farther, and the continual washing action

3

Figure 3.5 Coastal effects: A, raised beach level backed by old sea cliff, Lismore, Argyllshire, Scotland. IPR/73–34C British Geological Survey. © NERC. All rights reserved. B, ripple marks on beach sand produced by tidal currents; C, sun cracks in dried-out river mud.

of the water results in the concentration of sand in one part of the beach and pebbles in another. This sorting effect is also observed in certain sedimentary rocks and provides clues as to their origin. Structures seen on beaches include **ripple marks** (Figure 3.5B), **mud cracks** (Figure 3.5C) and **rain prints**, all of which can also be seen on bedding surfaces of sedimentary rocks. Ripple marks are formed by a current and typically have the shape of an asymmetric wave, with the steeper slope on the down-current side of the ripple. Ripple marks formed by tidal action show alternate asymmetry, that is, some beds show ripples formed by a flow tide and others, with the opposite sense of asymmetry, show ripples formed by an ebb tide. This feature enables tidal ripples to be distinguished in the geological record from current ripples, which usually form in deeper water.

Unsorted sediment results from the action of fast-flowing currents called **turbidity currents**. These operate intermittently, notably in deeper water, where they are triggered by sudden instabilities, caused by earthquakes for example, and result in turbulent flow of already deposited sediment from the continental shelf down the continental slope. These flows can cut quite deep underwater valleys known as **submarine canyons** and result in the deposition of layers of unsorted sediment characterised by upwards gradation from coarser to finer. These are known as **turbidites**, and have been used as an indicator of deep-water conditions (although shallow-water examples also occur). Away from all land-derived sediment, the ocean floor is covered by a thin layer of very fine mud formed mainly of volcanic dust and the remains of tiny oceanic organisms.

The desert environment

Desert landscapes show significant differences compared with those of temperate or tropical regions where the role of water is so important in erosion and transport, and in maintaining a cover of vegetation. Deserts are characterised by extreme dryness; rainfall is restricted to occasional short periods, causing flash floods, or may be completely absent. The main agent of erosion and transport in such environments is wind. Severe dust storms can carry small particles for huge distances and the larger sand grains can be rolled or bounced along the ground. Wind action can thus produce a size sorting very similar to that found in water-laid deposits, by blowing the mud and sand away, leaving the larger pebbles and boulders. In typical desert terrain, therefore, the rock outcrops are mantled by coarse debris aprons while the sand has been blown into dunes. The dunes are continually changing their shape and position because of the effect of the wind rolling the sand up the up-wind side of the dune and depositing it on the down-wind side.

Particles carried by the wind have a sand-blasting action on exposed rock surfaces, causing them to become polished; this action also has the effect of rounding the grains to the extent that many become completely spherical, with surfaces having the appearance of frosted glass. Such particles are called **millet-seed** grains, and if found in sedimentary rocks are a good indicator of past desert conditions. Small stones can become sand-blasted, producing characteristic faceted shapes.

Because of the lack of vegetation cover, when the occasional rainfall does occur, run-off is rapid, and the resulting flash floods

are important agents of erosion and transport in the comparatively short periods in which they operate. These floods typically occupy steep-sided, flat-bottomed valleys, called **canyons** or **wadis** (Figure 3.6A) and deposit vast loads of coarse detritus in debris fans at the mouths of the canyons. Finer sediment may be carried into short-lived lakes that soon dry out leaving sun-cracked mud. Salt and other soluble minerals may be precipitated from these lakes to form what are known as **evaporite** deposits (*see* Chapter 4).

Rock outcrops in desert regions tend to form steep sides and flat or rounded tops, caused by the sand-blasting action wearing away the base of the cliffs, rather like the action of waves on sea cliffs. In regions of horizontal **strata**, erosion of this kind forms the characteristic table-like landforms called **mesas** (Figure 3.6B). Continual wearing away of these structures eventually leads to the isolated pillar-like hills termed **buttes**, which are so characteristic of the desert states of the USA and which form the backdrop to many a Western film.

The effects of ice

Although glaciers disappeared from the landscape of the British Isles about 12,000 years ago, their effects can still be seen in all the main mountain regions of the country. However, to understand the processes of glaciation, and how the glacial landforms were produced, one has to look to the great ice sheets of Greenland and Antarctica, or the valley glaciers of the Alps or the Rocky Mountains, for example. The large continental ice sheets, like the one covering the greater part of Greenland, are over 3 km in thickness in places, and the great weight of this ice mass has depressed the land surface in the

Figure 3.6 Desert erosion: A, canyon, Bryce Canyon, Utah, USA; B, mesa and buttes in desert landscape, Arizona, USA.

central part of the sheet. When these ice sheets are removed, the drainage system becomes modified so as to flow inwards rather than towards the coast, and a landscape dominated by lakes emerges, such as exists today in the northern parts of Canada and Scandinavia.

These large ice sheets are regions of accumulation of ice through snowfall, but there is also a continuous outward flow of ice due to gravitational pressure. Near the margins of the sheet, where there is a suitable outwards slope, or pre-existing valleys exist that can direct the flow towards the coast, lobes of ice will move downwards to form valley glaciers (Figure 3.7A). Both ice sheets and valley glaciers continually gain ice mass from snowfall and lose mass by melting and evaporation at the margins. The ice crystals in snow are loosely packed (snow is ~90% air). However, the weight of overlying snow squeezes out the air to produce solid ice. If the system records a net gain, the valley glaciers will advance, and retreat if it suffers a net loss, but the ice flow continues regardless of whether the glaciers are advancing or retreating.

Valley glaciers are systems of moving ice similar to river systems in that they act as both erosive agents and transporters of eroded debris. Ice of course moves much more slowly than water, but varies considerably in speed, from under 1 m per year to over 100 m per day in exceptional cases. A typical Alpine glacier moves at around 40 m in a year. Ice movement is due partly to flow in the solid state, and partly to sliding over a film of water at the base of the ice sheet. Even in the coldest regions, the pressure of overlying ice is able to lower the melting point of the ice at the base of the sheet sufficiently to cause limited melting. Flow in the solid state takes place by means of

gradual changes in the shapes of individual ice crystals and sliding of crystals over each other under the influence of gravitational pressure.

The upper parts of glaciers are usually covered by snow but the lower parts show many signs of the movement of the ice. Deep cracks, or **crevasses** (Figure 3.7B, C), form in the upper brittle parts of the glacier as it is stretched over downward-bending sections of the floor; these are bent laterally into curved shapes as the glacier proceeds downwards because of the higher rate of flow in the central part of the glacier compared to the walls, where friction slows down the flow rate.

Although the rate of movement is so slow, the erosive power of a glacier is much greater than that of a river, partly because of its sheer size, but mainly because the ice carries considerable quantities of rock debris; this enables it to act like a piece of sandpaper, scraping and polishing the rock over which it passes. One of the most spectacular features of glaciers is the long ribbons of rock debris on the glacier surface. These initially mark the position of debris aprons along the margin of the glacier; however where two branches meet, one of the ribbons will now be in the centre of the combined glacier, and several parallel ribbons can be formed in this way as a result of branches joining the main ice stream.

The rock debris on a glacier surface includes all sizes of material from fine dust to large boulders; there is no sorting effect like that which operates in a river system. Part of the rock debris sinks through the ice, and part is carried to the front of the glacier and deposited there. Debris is also excavated from the floor of the glacier and carried along within the lower part of the ice mass. When the glacier retreats upstream, or disappears, the ground

is littered with piles of unsorted rock fragments; this material is called **moraine** (Figure 3.7A). Large boulders are often left lying on the ground surface where they have been dropped by the ice. These are termed **glacial erratics** (Figure 3.8A), and when composed of a distinctive rock type far from its source, can yield valuable information about the direction of the ice movement.

Glacial moraine forms several distinctive types of landform which are diagnostic features of ice transport and may be recognised in any formerly glaciated terrain, even when heavily vegetated, such as the Scottish Highlands. **Lateral moraines** are ridges formed at the margins of the glacier and **terminal moraines** are formed at the ends (Figure 3.7A); successive positions of a retreating glacier may be marked by several terminal moraines. **Eskers** are long sinuous ridges marking the course of a sub-glacial river. It may seem strange at first that a ridge should mark a river, but the edges of the sediment were contained within a channel of ice which has now disappeared, and the deposited material is left standing above the surrounding ground level. Other features are formed in areas of stagnant ice covered by sheets of moraine beyond the glacier margin; the melting of this ice leads to an irregular topography marked by isolated moraine mounds, known as **kames**.

The erosive power of ice is demonstrated by the scale of the valley features that have been carved out by glaciers. Originally V-shaped

3

3

Figure 3.8 Glaciation: A, glacial erratic (note geologist for scale). IPR/73–34C British Geological Survey. © NERC. All rights reserved. B, glaciated corrie, heavily ice-scoured, with hanging valley, the Cuillins, Skye, Scotland. IPR/73–34C British Geological Survey. © NERC. All rights reserved. C, Ben Starav, Glen Etive, Scotland, from its northern flank, showing the three narrow rocky ridges leading to the summit enclosing two glacier-formed corries.

valleys are transformed into broad U-shaped troughs, cutting down from the original floor by as much as several hundreds of metres, and truncating the ends of side valleys to form **hanging valleys**. The **cirque** or **corrie** (Figure 3.8B) is another characteristic feature of a glaciated terrain. These are rounded hollows cut into the side of a mountain by the upper part of a valley glacier and are produced by the plucking action of the ice as it attempts to flow downwards away from the rock onto which it is frozen. A mountain cut into from several directions by corries, is left with a sharp central peak surrounded by ridges, typical of the mountain scenery of northern Scotland, for example (Figure 3.8C).

The enormous scale of the excavation required to explain these features might seem out of proportion to the size of the glacier involved, but it must be remembered that even a small glacier moving at a rate of 100 m per year will have moved the equivalent of 1000 km of ice in 10,000 years, and that a sharp-edged block of hard rock frozen into the base of a thick sheet of ice under pressure is a remarkably effective cutting tool.

The abrasive effects of glaciation are preserved in glaciated landscapes by the presence of abundant grooves, termed **glacial striations**, and by smoothed and polished rock surfaces. Glacial striations are useful clues to the direction of ice travel. Other signs are asymmetric rock mounds with their steep sides facing downstream. These are called **roches moutonnées** and are caused by the ice plucking off blocks of rock on the downstream side of the mound by freeze–thaw action as the ice moves forwards.

Lowland areas surrounding glaciated mountains are often covered by a thick sheet of glacially derived material termed **boulder clay** or **till**, which consists of fine clay containing unsorted, angular rock fragments. This till may be moulded by moving ice sheets into long smooth rounded hills called **drumlins**, which are aligned parallel to the direction of movement of the ice sheet.

Glacial deposits

3

Evidence of former glaciations deduced from the geological record relies on the characteristic features of glacial-derived deposits. As just explained, these deposits are unsorted and contain angular blocks, contrasting with the well-sorted sands and rounded pebbles found in deposits from river systems. Larger blocks may display evidence of glacial striations or polishing. However, all glacial systems terminate in river systems or in the sea, and there is inevitably some mixture of effects in the final product. For example, angular blocks from glacial moraine can become rounded and partly sorted by river action.

However, lakes and shallow seas in glaciated regions do exhibit characteristics that are not shared by solely fluviatile systems. Lakes that are frozen over in winter receive a layer of sediment during the spring thaw, but the finest mud takes longer to settle and forms a fine layer on top of the coarse material. Thus a sequence of annual layers is built up, termed **varves**, that are useful in providing a record for dating purposes. Bodies of water overlain by floating ice accumulate very fine mud that contains scattered stones (**drop-stones**) and larger angular blocks that have dropped from the overlying ice. Such deposits are diagnostic of former glaciations (*see* Figure 4.4A).

Mud, sand and other deposits

We saw in the last chapter how the processes of erosion broke down rocks to form sediments, which were then carried by water, wind or ice to be deposited elsewhere. Such deposits form layers of sediment that may extend over large areas, particularly in the lower parts of river systems (on flood plains, for example) or in lakes, or in the sea. In general, each new layer covers a previously deposited layer, and thus younger **sediments** lie above older.

Over long periods of time, thick sequences of such sediment layers can build up, provided that there is a continued supply of sediment from the source region, and that the receiving area can accommodate that supply. Thick sedimentary sequences of water-lain deposits can only form where they can be laid down below water level, which means that in shallow water the area of deposition would need to sink progressively through time, relative to the water level, to accommodate them. Obviously, greater thicknesses of sediment can be accommodated in deeper water without the need for any depression. When a layer of sediment is buried beneath a sufficient thickness of younger material, it becomes transformed by a combination of physical and chemical processes into a sedimentary rock. These processes will be discussed later.

The sedimentary record

Sedimentary rocks are usually arranged in layers, termed **beds**, which vary in thickness from a few millimetres to many metres (Figure 4.1A). Thinner layers, down to about 1 mm, are termed **laminations**. Such layers represent a more or less continuous influx of sediment of a particular type, such as sand or mud, and are separated from the layers above and below by a **bedding plane**, which represents a change in sediment supply. This change could be a pause in supply, or a switch from one type of sediment to another. We noted in the last chapter how the laminated (varved) clays that form in glacial lakes represent seasonal variations in sediment supply. Other bedding planes might represent sudden changes in the sediment supply caused by floods or droughts, or differences in the source rocks accessed by erosion. Groups of beds of similar type are known as **formations**. Major, long-term, changes in sediment type mark the boundaries between adjacent formations, which can be easily distinguished from each other by the nature of their beds. For example, one formation may consist mainly of limestone beds alternating with mudstones, while another may consist mainly of sandstones and mudstones. Individual beds each represent a separate geological event, which may be of very short duration (a flood, say) or very long duration (many years of steady, continuous deposition) and vary widely in extent. Some sandstone beds extend for only a few hundred metres while beds of mudstone or limestone usually cover a much larger area; however, all beds eventually die out laterally, or merge into beds of a different type.

The Grand Canyon in Arizona (Figure 4.1B) is one of the best-known examples of a sequence of sedimentary rocks. The sides of the canyon expose a vertical thickness of almost 2000 metres of horizontal beds divided into seven separate formations, representing 300 million

Figure 4.1 Sedimentary rock layering: A. **Bedding (strata)**: alternate layers of harder and softer sedimentary rock create distinct ledges at the sides of this canyon in Arizona, USA.

4

B. **Formations**: the Grand Canyon cuts through a 2 kilometre-thick sequence of horizontally bedded sedimentary rocks grouped into several formations; the white cliffs are carved out of limestone formations; the vegetated slopes are in the more easily eroded shale formations, Arizona, USA.

years of deposition. Those formations consisting of more resistant beds of limestone and sandstone form escarpments while the more easily eroded shale formations form sloping terraces. The topmost bed is also the youngest, and was formed below sea level, from which we can deduce that the area must have been uplifted by at least 2 km since these beds were deposited. Such episodes of uplift in eroding areas, combined with corresponding episodes of depression in the receiving areas, are necessary to explain how the processes of erosion and deposition have continued throughout the enormous span of geological time.

The succession of sedimentary beds and formations is the basis of the geological record, which can be likened to a book, whose pages lie in sequence one upon another, each page representing the laying down of a layer of sediment and recording an episode of Earth history. The study of Earth history is called **stratigraphy**, one of the founders being William Smith, who, in the late 18th century, was able to recognise individual sedimentary formations by the fossils they contained, and to map them over large areas of the English countryside during his work on canal construction.

Types of sedimentary rock

There are three main types of sedimentary rock (Table 4.1): **clastic**, composed of fragments eroded from older rock; **chemical**, formed by chemical processes, and **organic**, formed either from organisms or from organic processes. Many rocks are mixtures of two or all three of these types.

Clastic rocks are divided according to the size of the dominant clasts in the rock, but also to an extent by the degree of roundness

clastic	coarse	conglomerate, breccia
	medium	sandstone
	fine	siltstone
	very fine	mudstone, shale
chemical		limestone
		gypsum
		rock salt
organic		coal
		ironstone
		limestone

Table 4.1 Some common types of sedimentary rock.

or angularity of the clasts. Sedimentary rocks with a high degree of sorting will show a uniform clast size but those that are poorly sorted may contain clasts ranging from mud to sand or sand to pebbles, and may be less easy to classify (Figure 4.2).

Rocks dominated by clasts larger than 2 mm in diameter are termed **conglomerate** where the clasts are rounded, but **breccia** where they are angular. Conglomerates with very large clasts may be referred to as **cobble conglomerates** or **boulder conglomerates**.

Rocks with clasts between 0.06 mm and 2 mm in diameter are classified as **sandstone**, but where such rocks are composed of angular fragments, they are often known as **grit** or **gritstone**. A special type of poorly-sorted sandstone, characterised by angular fragments of rock in a matrix of mud or sand, is called a **greywacke** and is produced by rapidly flowing, turbulent currents carrying a mixture of particle sizes in suspension (*see* below).

Finer-grained rocks with clasts between 0.06 mm and 0.004 mm in diameter are termed

4

Figure 4.2 Clastic sedimentary rock: rounded water-worn pebbles lie in beds of coarse-grained sandstone; the large pebble is about 6 cm across.

4

siltstone; and those with clasts less than 0.004 mm as **mudstone**. Laminated mudstones are known as **shale**. It is not necessary to actually measure the clast size to classify the rock. Mudstones and shales feel smooth to the touch and the grains cannot be seen except under the microscope. Silt grains may be seen with the aid of a hand lens, while sand grains are easily visible to the naked eye, so that classification is less complex than might appear!

Most chemical sediments are produced by precipitation from salty water due to the evaporation of shallow seas or lakes in hot climates, and are termed **evaporites**. They include some **limestones** (calcium carbonate), **gypsum** (calcium sulphate) and **rock salt** (sodium chloride). Gypsum may be more familiar in the form of **alabaster**, which is much prized as an ornamental stone because of its relative softness and ease of working. Typically it is shiny white in appearance with purplish or reddish veining.

Organic sediments are formed wholly or mainly from the remains of fossil organisms, or as a result of organic processes; they include some limestones, coal, some ironstones and some cherts. **Limestone** is composed largely of calcite (calcium carbonate) and to a lesser extent **dolomite** (calcium magnesium carbonate) which in most cases is of organic origin – derived from either whole fossils or fossil fragments. **Coal** is formed from plant debris. There are various types of **ironstone**; a common variety consists of a mixture of silica and iron oxide (**hematite**) and is typically reddish brown in colour. Ironstones are formed by precipitation under certain conditions, which may be aided by the activities of algae or bacteria, and thus may be partly chemical and partly organic in origin. **Chert** is composed of microcrystalline silica and is extremely hard and dense. **Bedded chert** is formed by the accumulation of microscopic organisms whose skeletons are made of silica, and is usually black, green or reddish-brown in colour (*see* Figure 4.4B). It is formed in very deep water, far from any source of clastic sediment. Organic or chemical sediments such as limestones may be eroded and re-deposited as clastic sediments.

From mud to rock

Soft sediment is converted into hard rock by means of a number of processes that take place when the sediment has been buried to some depth beneath younger deposits. The sediment is compressed by the pressure of overlying material, which pushes the grains closer together, squeezing out the air and water, and resulting in a more compact and denser material. Muds compress to about one fifth of their original volume, and sands to about nine-tenths. Changes also occur in the mineral contents. For example, sand grains become cemented together by minerals deposited from solution. The commonest of these are quartz, calcite and hematite (iron oxide). The presence of hematite **cement** is very obvious in sandstones because of the red or brown colour it gives to the rock (e.g. Figures 4.2 and 4.3A). These cements make the rock very hard and resistant to erosion, especially where silica is the cement in a quartz sandstone; this results in a particularly hard rock termed a **quartzite**. The high degree of compaction experienced by muds results in many cases in a rotation of the minute flakes of clay minerals so that they lie parallel to the bedding plane. The resulting finely laminated rock is a **shale**.

Another type of change that takes place in buried sediments is the replacement or partial replacement of the existing minerals by new minerals deposited by percolating aqueous (water-rich) solutions. The commonest examples occur in limestones, where the original calcite may be recrystallised or replaced either by dolomite (calcium magnesium carbonate) or by silica in the form of **chert**. Hard nodules of grey **flint** (a type of chert) are widespread in the **chalk** (limestone) formations of southern and eastern England and were much used by our Stone-Age ancestors for tool-making.

Tracks and signs

At first glance, one sandstone or mudstone may look much like another, whether it was formed in a river bed or an estuary or in the deep sea. However, closer inspection may reveal clues about the original environment of deposition. Such clues are of two kinds: those that throw light on the physical processes accompanying deposition, flowing currents for example, and those that reveal the type of organic life that existed during deposition.

Bedding has already been described; the scale of the bedding (whether finely or thickly bedded) is an important indication of the quantity of sediment deposited during any particular depositional episode. **Cross-bedding** (Figure 4.3A) describes a structure where one set of beds is deposited on an inclined slope while another set, which may cut across earlier sets, is deposited at an angle to it, either horizontally, or inclined in a different direction. In this type of structure, the younger beds always cut across the older. Different types of cross-bedding form in different environments. **Dune bedding** is a large-scale type of cross-bedding formed in sand dunes, where successive layers of sand are deposited on the lee slope of a dune. The cross-bedding formed

4

Figure 4.3 **Sedimentary structures**:
A, **cross-bedding**: the lower set of beds inclined to the left are cut off by the upper set inclined to the right; the structure is caused by a change in the direction of flow of the current. B, **mud cracks**: this displaced slab of rock shows the underside of a sandstone bed; the ridges result from infilling of cracks in the underlying mud. C, **load casts in graded bedding**: the coarser silt (now siltstone) above has caused shallow depressions in the underlying mud (now mudstone) which has been squeezed up into the silt in the form of narrow 'flame structures'.

in estuaries and deltas is usually on a smaller scale. Here the inclined layers are deposited on the slopes of sand bars or at the ends of deltas and usually have shallower inclinations than in dune bedding.

Current ripples are formed on the surface of a bed; they are asymmetric, with the steeper slope on the down-current side. Successive ripples migrate in the direction of the current. Tidal currents produce alternating asymmetry in the ripples because of the ebb and flow of the tides. Such structures are good indicators of shallow water and a near-shore environment. **Mud cracks** (Figure 4.3B) and **rain prints** are formed when shallow water dries out, and are restricted to shallow coastal waters and inland lakes.

Graded bedding (Figure 4.3C), where the coarser particles are concentrated at the base of the bed and the finer at the top, is an indication of a lack of sorting and is typical of deposition from **turbidity currents**. These currents are characterised by rapid, turbulent flow and are capable of carrying large loads of sediment of varying particle size in suspension. Currents of this type are intermittent in nature and occur where large thicknesses of soft sediment are poised above a slope where a sudden shock, such as an earthquake, may cause the sediment to slide down-slope like an underwater avalanche. As the flow slows down, the coarser, heavier particles settle quickly to the bottom while the finer settle more slowly to produce the graded bed. Turbidity currents may form in water of any depth, but are especially important sources of deep-ocean deposits, sending masses of mixed-up sediment in a fast-moving flow from the continental shelf down the continental slope into the deep ocean. Beds deposited from turbidity currents often show scour marks and grooves on the mudstone tops of the graded beds caused by the ploughing action of the coarse material destined to form the next layer as it is carried along by the fast moving current. These structures are more usually seen 'in reverse' as casts in the sandstone base of the bed above, which has infilled the scour marks. Such marks are useful in recording the direction of current flow.

The most useful clues to the depositional environment are in most cases organic in nature, either directly in the form of fossils of once-living organisms, or indirectly from the tracks or traces of such organisms. Thus plant remains are indicators of freshwater deposits (rivers, lakes, estuaries) while fossils with carbonate shells are indicators of marine conditions. Certain faunas are characteristic of deep waters and others of shallow, near-shore environments. There is a danger, of course, of circular arguments here, in that the mode of life of fossils in very old rocks, from groups unrepresented at the present day, may be a matter of speculation. However, there is usually enough evidence, taking the fossil assemblage as a whole, and combining it with the physical evidence, to determine the likely depositional environment of a fossil-bearing sedimentary rock. Traces of organisms can be equally useful; these include such clues as the footprints of land reptiles and worm burrows in shallow marine sediments.

Sedimentary environments

An important aim for the geologist is to piece together information about the sedimentary environments of rock formations of the same age in different places, and in this way to reconstruct the former geography (**palaeo-geography**) of that formation. Climatic patterns change with time, and the continents move around the globe, so that geographical reconstructions are essential in recording such changes and in providing evidence for continental drift.

At any given time, sedimentary deposits will be forming in a number of different environmental regimes; these can be divided firstly into continental and marine. The continental deposits include those formed in rivers and lakes, or on land, as in desert sand dunes. In the lower reaches of river systems, the estuarine and deltaic environments span the transition between freshwater and marine conditions. Marine environments range from coastal and near-shore, through continental shelf and continental slope, to the deep ocean. A complete picture of the palaeogeography of the world during a particular geological period would be made by finding examples of all these environments and making a map of their distribution.

There are a number of key indicators in making such reconstructions:

● **Red beds** (conglomerates, sandstones or shales) are restricted to continental or shallow marine environments in which there

Figure 4.4 A, **boulder tillite** formed by boulders dropped from floating ice into glacial mud. B, red and green **bedded cherts**.

4

is sufficient oxygen available to form hematite, the more highly oxidised form of iron ore.

● *Plant remains* are typical of freshwater lakes, and of estuarine and deltaic environments, but are rarely carried further out to sea.

● *Well-sorted deposits* are typical of near-shore sediments affected by constant wave and current action, and range from conglomerates through sandstones to mudstones; the sandstones and mudstones will usually show evidence of burrowing organisms, if they do not contain the fossils themselves.

● *Limestones* are typical products of the shallow seas of the continental shelf, although they can also form in freshwater lakes; the presence of shelly fossils, corals and other organisms would confirm their marine origin. Reefs are a special type of limestone deposit, characterised by the lack of an obvious bedded structure and by an abundance of reef-building fossils, such as corals.

● *Clastic sediments* are generally finer-grained the further they are from their source, and so mudstones are typical of the continental shelf and of the deep ocean floor. In addition to the fine mud normally deposited there, the continental slope and ocean depths receive occasional floods of clastic debris in the form of greywackes washed down by turbidity currents from the shelf.

● *Mudstones with scattered pebbles* ('dropstones') or boulders are a diagnostic feature of glaciated terrains, the large clasts having been dropped into the mud from floating ice (Figure 4.4A).

● *Bedded cherts* (Figure 4.4B) composed largely of silica deposited by minute organisms are typical of the very deep ocean floor, far from any source of clastic sediment.

Moving continents and making mountains

Most, if not all, readers will be familiar with the term 'plate tectonics'. This theory, first advanced in 1967, revolutionised the science of geology in much the same way as atomic theory transformed the science of chemistry at the beginning of the 20th century, in that it provided a new framework within which many previously unrelated or unexplained facts could be brought together and made sense of. This new theory was itself an amalgamation of two previous theories, **continental drift** and **sea-floor spreading**, both of which presented a 'mobilistic' view of the Earth's crust, in which individual continents and pieces of ocean crust were thought to be continually moving around the Earth's surface relative to each other, creating mountain belts in zones of convergence.

Continental drift

Continental drift is a comparatively old idea, first popularised by Alfred Wegener in 1915, which was proposed to explain geometric and geological similarities between continents now separated by oceans. The continents of South America, Africa, India, Australia and Antarctica were shown to fit together in a supercontinent called **Gondwana** (Figure 5.1), and North America and Eurasia fitted together into a second supercontinent called **Laurasia**. These two supercontinents appeared to be joined together in central America, forming a continuous worldwide landmass termed **Pangaea** (pronounced 'Pan-jee-a') enclosing the **Tethys Ocean** (Figure 5.2). It should be noted here that the term '**continent**' used in a geological sense includes, in addition to the landmass, areas of adjacent sea bed underlain by continental-type crust – the **continental shelf** and **continental slope**. When these are included, a much better fit of the Gondwana continents is achieved.

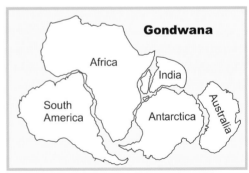

Figure 5.1 This arrangement of the five southern continents 200 million years ago is known as **Gondwana**. Note how well the coastlines fit together.

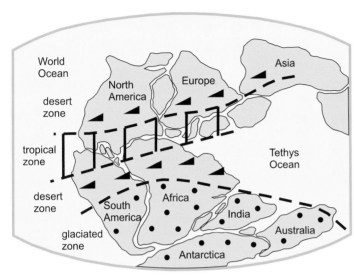

Figure 5.2 Climatic zones of the supercontinent of **Pangaea** 200 million years ago. These climatic zones form bands on either side of the equator of that period, which runs through southern North America and Europe, and the polar ice sheet covers large parts of all the present southern continents. Therefore the climatic zones make sense in the Pangaea reconstruction but not when the continents are in their present positions. The tropical zone is defined by coal deposits and coral reefs, the desert zone by dune-bedded sandstones and evaporite deposits, and the glaciated zone by tillites and glacial striations. Based on Hamblin (1989) figure 17.6, after Wegener (1929).

5

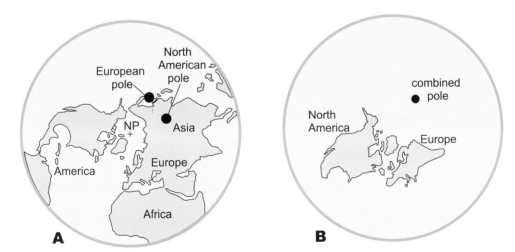

Figure 5.3 200 million-year-old north pole positions. A. Polar projection of the northern hemisphere showing the different positions of the north magnetic pole for 200 million-year-old rocks from Europe and North America. B. When the continents of North America and Europe are fitted together in the positions occupied in the Pangaea reconstruction, the magnetic north poles coincide. Based on McElhinny (1973).

When this continental reconstruction was examined, many geological features shared by the separated continents could be explained. For example, similarities in fossil land animals and plants that existed 200 million years ago prior to the splitting up of the supercontinents contrast with the obvious differences that exist now. Another telling piece of evidence is the presence of glacier-derived clays and striations in rocks of Carboniferous age in all the Gondwana continents, covering in their present positions about half the globe, but when restored to their presumed Gondwana fit, making a reasonably-sized polar ice cap. The distribution of other climatic indicators in rocks 200 million years old also makes sense when in the Gondwana fit; these include dune-bedded sandstones and evaporite deposits, which mark out two desert belts on either side of a central equatorial belt indicated by the presence of coal deposits and coral reefs, indicators of tropical conditions. The distribution of these climatic indicator rocks makes no sense in their present locations; for example, coals representing the product of equatorial forests now lie near the north pole, and glacial deposits lie near the equator!

Wegener's ideas caused considerable debate among the geological community, failing to obtain universal acceptance mainly because of the lack of a convincing mechanism for the movements. Physicists, in particular, opposed continental drift because their calculations of the strength of the Earth's crust 'proved' that it was incapable of the type of behaviour required. However, work on radioactivity led Arthur Holmes in 1931 to demonstrate that the Earth must be much hotter, and therefore much weaker, than previously thought, and suggested that the mantle could be capable of transferring heat by slow flow by means of **convection currents** in the solid state. Such mantle currents could carry continents laterally across the Earth's surface. Debate carried on, however, until the 1960s when work on **palaeomagnetism** (the magnetic directions of old rocks) showed that the positions of magnetic north for 200 million-year-old rocks in different continents plotted in different places. However, when the continents were fitted together in their presumed original positions, the locations of magnetic north poles coincided (Figure 5.3). This was convincing proof that the continents had drifted to their present positions from their supercontinent positions 200 million years ago.

The ocean floor – static or mobile?

The next stage in the evolution of ideas came from studies of the ocean floor, where mapping by various remote-sensing techniques had revealed a topography that was as varied as that of the continents. The generally even ocean floor is interrupted by a system of great ridges and deep, narrow trenches (Figure 5.4). The ridges are typically between 1000 and 2000 km wide and rise to as much as 3 km from the ocean floor. They make a continuous network, one branch of which runs from the Arctic along the centre of the Atlantic Ocean (the mid-Atlantic ridge) to join a second branch that completely surrounds Antarctica and crosses the Pacific Ocean towards the coast of Mexico, sending two branches into the Indian Ocean. The trenches are much narrower (typically around 100 km wide) but extend to depths of up to 11 km below sea level. They form generally curved linear features on the ocean-ward side of island chains around

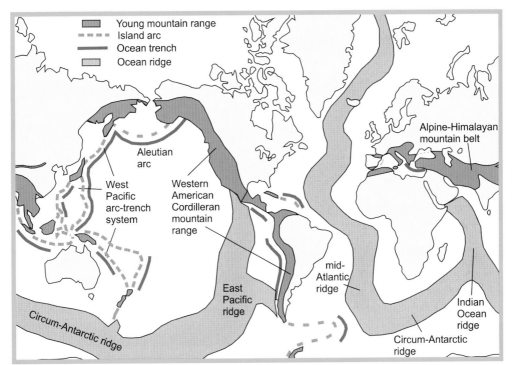

Figure 5.4 Main topographic features of the continents and oceans. The **young mountain ranges** form two distinct belts, the Alpine-Himalayan and circum-Pacific, meeting in Indonesia. The circum-Pacific belt consists partly of mountain belts along the western margins of North and South America and partly of a series of submerged mountain ranges (**island arcs**) around the north and west sides of the Pacific Ocean and on the northeast side of the Indian Ocean. The oceanic features are of two kinds: the **ocean ridges**, which are almost completely submerged, and the much narrower **ocean trenches**, most of which lie parallel to the island arcs.

the western Pacific Ocean, the eastern Indian Ocean, in the Caribbean, and along the Pacific coast of America.

Much of the objection to Wegener's ideas on continental drift had centred on the failure to visualise how a continent could plough across static ocean crust. However, this objection was countered by the proposal that the ocean crust behaved like a giant conveyor belt, rising at the ridges and moving sideways towards the deep ocean trenches where it descended (Figure 5.5). In other words, both continents and oceans were mobile rather than static. Wegener had thought that the ocean ridges marked the lines of separation of the continents, but **palaeomagnetic** dating of the ocean floor of the Atlantic and Indian oceans in the 1960s showed that the ridges were the most recently formed parts, and that the ocean floor became older towards the continental margins.

The dating of the ocean crust relies on the fact that new crust formed along the ocean

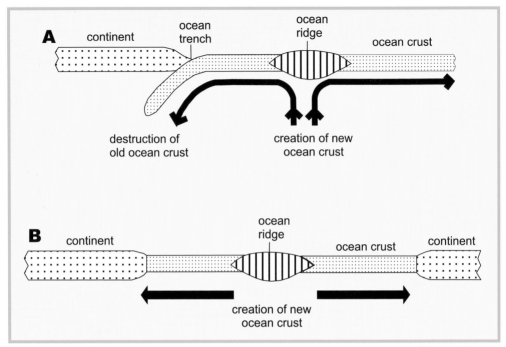

Figure 5.5 The 'conveyor-belt' model. How ocean crust is created and destroyed: A, creation and destruction of oceanic crust; B, how continents move apart by creation of oceanic crust.

ridges becomes imprinted with the contemporary magnetic field, and this changes periodically by swapping magnetic north and south poles. Each change creates a long strip of crust, parallel to the ridge axis, whose magnetic character differs from the previous one, and as new strips are created, older strips move away from the ridge axis. This process creates a series of strips (or magnetic stripes) on the ocean floor, each representing a particular period of formation. The stripe sequence can be calibrated with reference to dated lava flows on land.

This work proved that the continents of Gondwana and Laurasia had indeed moved apart and that the space between had been filled by new ocean crust, becoming younger towards the ridges, which were seen as the sites of formation of new ocean crust at the present day.

However, ocean crust could not be continuously created without it being destroyed elsewhere, and the obvious sites for destruction were the deep ocean trenches. The new palaeomagnetic dating evidence demonstrated that ocean crust adjacent to the trenches showed a variety of ages (a 'discordant' age pattern) whereas crust adjacent to continental margins that had moved apart showed a 'concordant' age pattern, the age being consistent with the date of separation of the continent. This evidence confirmed that the conveyor belt model for the ocean floor was essentially correct.

The plate tectonic model

The final stage in the construction of the plate tectonic model was based on the proposition that both the continental and the oceanic crust must behave in a semi-rigid manner, moving laterally as single units or blocks, and that relative movement between the blocks must be concentrated at their boundaries. This proposition arose from the observation that linear features on the ocean floor, such as faults and the striped magnetic pattern, were essentially unaffected by warping or bending such as might be expected if the ocean floor were to behave in a 'plastic' manner. The opposing coastlines of Africa and South-Central America can still be fitted together despite having travelled away from each other for a distance of 3000 km over a period of 150 million years, indicating that South America and the western half of the South Atlantic on the one hand, and Africa and the eastern half of the South Atlantic on the other could be considered as separate blocks, which moved as units.

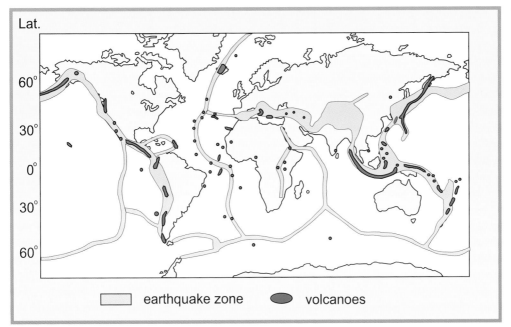

Figure 5.6 Pattern of recent earthquake and volcanic activity. The earthquakes follow well-defined narrow zones along the centres of the ocean ridges and rather broader zones along the Alpine-Himalayan mountain ranges and the circum-Pacific island-arc network. Most of the volcanoes lie within these same zones but some oceanic volcanoes are situated away from the ridge crests, especially in the Atlantic ocean.

It was then realised that the linear zones of earthquakes that follow the ridge–trench network (Figure 5.6) must be related to the movements taking place along the boundaries of the relatively stable blocks, and that the margins of these blocks could be mapped out by following the earthquake zones. The term **tectonic plate** was introduced to describe these blocks. The earthquake zones completely surround the stable plates, whose boundaries could now be seen to be of three types: ridges, trenches and faults (Figure 5.7). Moreover, since the ridges must mark sites of production of new ocean crust (**constructive boundaries**), and the trenches mark sites of destruction of ocean crust (**destructive boundaries**), the faults must correspond to zones where one block merely slides past its neighbour without either creation or destruction of crust taking place. The faults that form parts of the boundary network link constructive and destructive boundaries and are

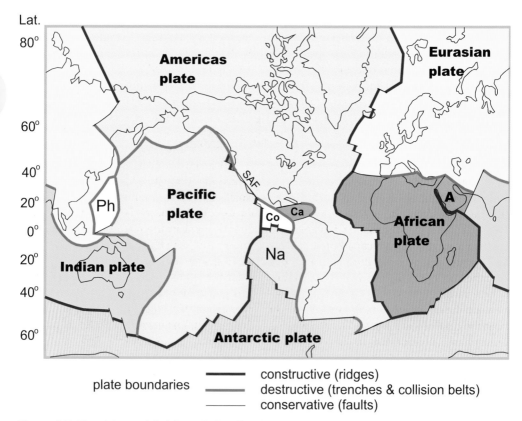

Figure 5.7 The plates and their boundaries. The plates are separated by three types of boundary: **constructive** – ocean ridges; **destructive** – ocean trenches and young mountain belts; and **conservative** – faults. Small plates: Na, Nazca; Co, Cocos; Ca, Caribbean; Ph, Philippine; A, Arabian. SAF, San Andreas fault.

termed **transform faults**, since they 'transform' one type of motion (e.g. convergent) to another (e.g. divergent). Because plate is 'conserved' at these boundaries, they are known as **conservative boundaries.**

The rates of movement of the plates can be estimated from the dating of the magnetic sea-floor stripes; these give rates of the order of centimetres per year. Such rates can be verified by present-day precise measurement, of the separation of Europe and America, for example. They seem very slow to us but, measured over geological time, are substantial, e.g. 100 km in one million years is not untypical. Plate movement at the surface, of course, is not continuous but consists of short phases of rapid movement, which generates earthquakes (*see* Chapter 6), separated by long periods of inactivity. These average out, over geological time, to the 'long-term' rates stated above.

The plates are pieces of a strong upper layer of the Earth, termed the **lithosphere,** consisting of the crust and the uppermost part of the mantle. This strong layer rests on a weaker zone, termed the **asthenosphere**, detectable by seismic waves, which is capable of very slow solid-state flow, allowing the plates to move over the underlying mantle. The oceanic lithosphere varies in thickness from around 50 km beneath the ocean ridges to more than 100 km near the continental margins. The continental lithosphere is considerably thicker.

Making new plate

The construction of new plate takes place along the central zones of the ocean ridges and within **continental rift** valleys such as the great **African Rift**. Some of the new material takes the form of basalt magma produced by melting of the upper mantle in the hot low-density region beneath these constructive boundaries (*see* Figure 2.7). This magma is injected into the crust in the form of basalt dykes or gabbro sills, and part is extruded onto the surface as lava flows. Where the lavas are poured out under water, as they are over most of the ocean ridges, they often form **pillow lavas** (*see* Chapter 2) which are a diagnostic feature of submarine flows. These basaltic rocks form the new oceanic crust; beneath it, new oceanic mantle lithosphere is formed by the addition of ultramafic material transferred by ductile flow from the asthenosphere and subsequently cooled.

The focus of present-day activity at these constructive boundaries is marked by earthquakes and volcanic activity in a zone around 100 km wide along the centre of the ocean ridges and within the continental rift valleys. Part of this volcanic activity is in the form of '**black smokers**'. These are hydrothermal vents which discharge superheated water containing dissolved minerals such as sulphides and deposit them in the form of black chimney-like structures rich in metallic ore deposits As new volcanic material is added to a ridge, the older parts move away from the central zone and gradually cool down, subsiding as they do so, until they reach the normal level of the deep ocean. The great African Rift is part of a rift system that includes the Red Sea and the Gulf of Aden (Figure 5.8). Both these structures contain strips of oceanic crust along their centres and may eventually become oceans if Arabia and Africa move apart. In this way the continents of the Americas, Europe and Africa would have separated during the split-up of Pangaea (*see* Figure 5.2).

5

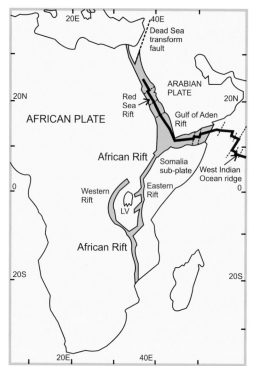

Figure 5.8 The **Red Sea–Gulf of Aden–African Rift** System. New oceanic crust is being formed along the Red Sea and Gulf of Aden rifts to join up with the West Indian Ocean ridge causing the Arabian plate to break away from the African plate. No separation has yet occurred along the African Rift part of the system although considerable volcanic activity has occurred here over the last 40 Ma or so.

Plate destruction

There are two types of destructive plate boundary. The first type follows a zone of destruction of oceanic crust and is marked at the present day by the deep-ocean trenches, the second type is a zone of collision of two continental plates, and is marked at present by a belt of young mountain ranges such as the Alps and the Himalayas. The second type

of boundary is, in geological terms, only temporary, since convergent movement will eventually cease as the two continental plates grind together and gradually come to a halt.

The standard case of the destructive boundary therefore is where oceanic crust on one plate (the lower) descends beneath another plate (the upper) which may be either continental or oceanic. The line of descent is marked by a deep-ocean trench and is known as a **subduction zone**; this zone is typically inclined beneath the upper plate (Figure 5.9A, B). As the lower plate descends into warmer regions of the mantle, some of the crustal material melts, and the resulting magmas ascend into the upper plate forming igneous intrusions within the upper-plate crust and volcanoes at the surface. Where this zone of volcanoes lies on an oceanic upper plate, it is partly submerged and forms an **island arc**. Present-day examples of continental-margin subduction zones are situated along the Pacific margins of South and Central America while volcanic island arcs are widely distributed in the western Pacific and east Indian oceans and in the Caribbean (*see* Figures 5.4 and 5.6).

In a typical island arc, the partially submerged volcanic mountain range, 50 to 100 km wide, is bounded on its convex side by a trench situated between 50 and 150 km from the volcanic arc. The zone between these two structures – the arc-trench gap, or **fore-arc** – is a zone of sedimentary accumulation resulting from the erosion of the volcanic islands. Some of this material eventually descends via **turbidity flows** (*see* Chapter 4) into the trench as sediments – such as **greywackes** and slide **breccias** – that can be recognised in the geological record and can be used to identify former subduction zones.

6

Earthquakes and faults

An **earthquake** is a vibration of the ground caused by a sudden displacement or failure of rock at depth. The vibrations caused by this event spread out in all directions from the source of the disturbance, like ripples in a pool when a stone is thrown in. Earthquakes are common and very widespread; most are too small to be detected except by sensitive instruments, but the largest ones cause immense damage and loss of life (Figure 6.1).

Figure 6.1 Earthquake damage, Izmit, Turkey, 1999. This 7.6 magnitude earthquake killed around 17,000 people and left half a million homeless. IPR/73–34C British Geological Survey.
©NERC. All rights reserved.

Intensity and magnitude: the 'size' of an earthquake

The destructive power of an earthquake – its **intensity** – depends on the severity of the ground motion, and is measured by the effects felt at the surface. Intensity is usually described in terms of numbers on the **Mercalli scale** (Table 6.1) and ranges from instrumental (detected only by instruments) to catastrophic (total destruction of all buildings). In populated areas, it is possible to map zones of increasing intensity towards a central position above the source of the earthquake where the intensity is at a maximum. This point is termed the **epicentre**. The **magnitude** of an earthquake is a more exact way of describing its size, and is usually measured on the **Richter scale** (Table 6.1). The magnitude measures the amount of energy released by the earthquake at its source and is reflected by the severity of the surface vibrations, or **earthquake waves**. The most severe earthquakes known have a magnitude of between 8 and 10 on the Richter scale whereas the weakest detectable by humans have a magnitude of about 3.5. Those weaker than this are only detectable by instruments. Each successive number on the Richter scale represents a

	Intensity	Characteristic Effects	Approximate equivalent magnitude
I	Instrumental	detected only by seismograph	1—3
II	Feeble	detected only by sensitive people	
III	Slight	felt by people at rest, like vibrations from a passing heavy vehicle	
IV	Moderate	felt by people walking; rocking of loose objects	4—5
V	Rather Strong	felt generally; people awakened; bells ring	
VI	Strong	trees sway; suspended objects swing; loose objects fall	
VII	Very Strong	general alarm; walls crack; plaster falls	
VIII	Destructive	drivers disturbed; masonry cracked, chimneys fall; damage to poorly constructed buildings	6—7
IX	Ruinous	some houses collapse; ground cracks; pipes break	
X	Disastrous	ground cracks badly; many buildings destroyed; railway lines bent; landslides on steep ground	
XI	Very Disastrous	few buildings remain standing; bridges destroyed; railways, pipes and cables out of action; great landslides and floods	
XII	Catastrophic	total destruction; objects thrown into the air; ground rises and falls in waves	8 +

Table 6.1 Characteristic effects of earthquakes of different intensities (**Mercalli scale**) and the approximate equivalent magnitudes (**Richter scale**).

factor of ten times more energy than the last; thus a magnitude 5 earthquake is ten times more severe than a magnitude 4, and so on.

The damage and loss of life caused by a major earthquake can be catastrophic, as demonstrated by two notable examples in the year 2004 – the Bam earthquake in Iran, and the Indonesian earthquake west of Sumatra. The Bam event caused widespread destruction to a whole city. In the Indonesian occurrence, the displacement of the ocean floor caused by the earthquake produced a huge ocean wave, or **tsunami**, that devastated vast areas of low-lying ground, not only in neighbouring

Indonesia and Thailand but also across the Indian Ocean in Sri Lanka. In this example, the destructive effects were due much more to the tsunami than to the earthquake vibrations themselves. A more recent devastating earthquake occurred in January 2010 in Haiti. The focus of this 7.0 magnitude earthquake lay at a shallow depth of only 13 km a mere 25 km west of the capital, Port au Prince, and caused widespread destruction; an estimated 3 million people were affected and tens of thousands lost their lives. The cause was movement on a transform fault separating the eastward-moving Caribbean plate from the westward-moving Americas plate (*see* Figure 5.7)

In general, destructive effects bear no simple relationship to the size of an earthquake and depend much more on the quality and construction methods of the buildings involved, and indeed whether the earthquake occurs in an inhabited region. In recent years severe earthquakes have occurred in Japan with much less consequential damage because the buildings there are constructed to withstand earthquake vibrations.

The causes of earthquakes

Earthquakes originate within the Earth's crust and uppermost mantle, where the rock is strongest and most liable to sudden failure. At greater depths the rocks are weaker and tend to flow rather than fracture. They are divided into **shallow focus** (originating at depths down to 70 km), **intermediate** (from 70 to 300 km) and **deep** (between 300 and 700 km). Shallow earthquakes are typical of ocean ridges and faults along constructive or conservative plate boundaries (*see* Figure 5.6). Intermediate and deep earthquakes, on the other hand, are typical of destructive plate boundaries where the plates

have been subducted to much greater depths. Although earthquakes are heavily concentrated along plate boundaries, and help to define them, shallow earthquakes also occur within plate interiors.

There are over one million earthquakes every year, but very few of these are large. Their frequency is inversely proportional to their magnitude: thus in a given year there might be about 700,000 earthquakes of magnitude 1 but only 20 of magnitude 8.

The amount of energy released every year by earthquake activity is probably constant. This energy results from the continual pressure exerted by the forces acting on the Earth's moving plates (*see* Figure 5.10) and is released when the pressure builds up to the point when it overcomes the strength of the rocks. This happens more often at plate boundaries, where relative plate motion is accompanied by large opposing forces, but because the pressures are transmitted through the plate interiors as well, any weakness, such as an old fault, can result in the release of part of this pressure in the form of an **intraplate** (within-plate) earthquake.

The initial rock failure that causes an earthquake may spread rapidly along a fracture (or fault) plane over distances of many tens of kilometres. The major San Andreas fault zone of California has experienced countless earthquakes along its 1200 kilometre length over a period of several million years, contributing to a total displacement estimated at several hundred kilometres.

Although rock failure is usually associated with the formation or re-activation of a fault in the rock, it can also occur as a result of volcanic activity, particularly when accompanied by the explosive release of gas.

6

Earthquake waves

The shock waves from an earthquake are recorded on an instrument called a **seismograph**, on which the ground vibrations are traced on a revolving drum in the form of a series of wave forms (Figure 6.2A). The first waves to arrive are termed **primary (P) waves**, and are followed by a second set, termed **secondary (S) waves**, followed in turn by a third, termed **surface waves**. Both primary and secondary sets take a 'short cut' through the Earth (Figure 6.2B) whereas the surface waves travel around the Earth's surface and thus take longer to arrive. The arrival time of the surface waves is directly related to the distance from the earthquake source measured around the surface.

The location of the source of an earthquake may be determined using the time difference between arrival times of the primary and secondary waves, which increases with distance from the source. A minimum of three seismographs situated in different directions from the source is necessary to find the epicentre, which is located at the intersection of the three circles with centres at each of the three seismograph stations and radii equal to their respective distances from the epicentre (Figure 6.2C).

The P and S waves from large earthquakes can be used to study the structure and physical properties of the Earth's interior. Their velocity is dependent upon the density of the rock through which they travel, and thus variations in density of the various Earth layers can be deduced by mapping out the arrival times at seismographs around the world (Figure 6.2B). No S-waves pass through the outer core, creating a circular **shadow zone** on the opposite side of the Earth to the earthquake (Figure 2.6). Their absence is explained by the fact that S-waves cannot pass through a liquid medium and demonstrates the liquid nature of the outer core. The position of the edge of the shadow zone can be used to locate the mantle-core boundary.

Earthquake prediction

Much research has been carried out in an attempt to predict major earthquakes accurately. This has been only partially successful. Whereas it is possible to forecast that a large earthquake is likely, there is as yet no method of predicting the exact timing of the event. Along large fault zones such as the **San Andreas fault zone** of Western USA, which marks a plate boundary, repeated earthquake events have occurred in the past and can confidently be predicted in the future, for as long as there is relative motion between the Americas plate and the Pacific plate to the west. Sections of the fault zone that have recently experienced movement are unlikely to fail again soon but may transmit pressure to adjoining sections. The longer a specific segment of the fault zone is dormant, the larger the earthquake event is likely to be when it occurs.

It is possible to monitor sections of a fault zone that lie within or near inhabited areas to detect the small changes in the ground that precede failure. The rocks contract or expand slightly due to increases or decreases in pressure. Pores and cracks may open up affecting the flow of fluids, and these changes are accompanied by changes in physical properties, such as electrical resistance. However, the exact time of the earthquake is impossible to predict with sufficient accuracy as yet, and of course failure might come in a part of the zone that has not been monitored.

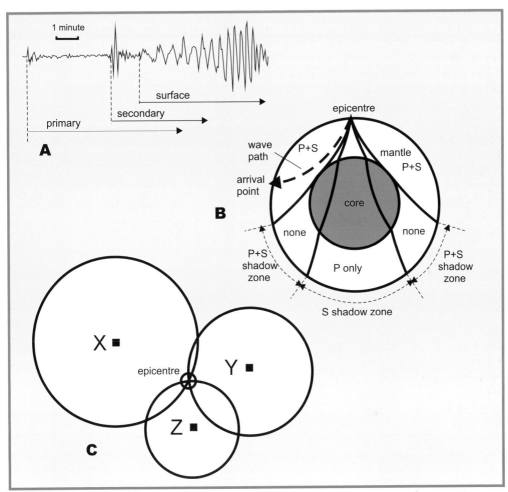

Figure 6.2 A. Typical earthquake-wave recording, showing **primary**, **secondary** and **surface wave** traces. B. Diagram of a cross-section through the Earth, showing a typical curved wave path and arrival point. Note the gap (**shadow zone**) between the paths of the deepest waves to pass entirely through the **mantle** and the shallowest (P-waves only) to pass through the **core**. The position of this ring-shaped (in 3D) shadow zone determines the size of the core. C. Method of locating the **epicentre** of an earthquake detected at recording stations X, Y and Z; the radii of the circles represent the respective distances from these stations to the epicentre.

Fractures and faults

Rock fractures are abundant and can be seen at any rock outcrop in the form of cracks interrupting the continuity of the rock surface.

Where there has been obvious movement on the crack, the fracture is termed a **fault**; where no displacement is visible, the fracture is termed a **joint**.

Types of fault

Faults are classified according to whether the movement occurs horizontally along the fault plane (**strike-slip faults**) or up and down the fault plane (**dip-slip faults**) (Figure 6.3). Those where movement is oblique are termed **oblique-slip faults**. Dip-slip faults are normally inclined, with upper and lower sides (Figure 6.3A). Where the upper block has moved downwards on the fault plane, the structure is termed a **normal fault**; in this case the rock has been extended. Where the upper block has moved upwards on the fault plane the structure is termed a **reverse fault** and the rock has been compressed (Figure 6.3B). Normal faults are usually steeply inclined and are much more common than reverse faults.

Thrust faults and crustal shortening

A special category of reverse fault is termed a **thrust fault** and is typically gently inclined or even horizontal. Thrust faults play an important role in mountain belts by transferring rocks from deep crustal levels up to the surface by moving them over younger, originally higher-level, rocks (Figure 6.4). In this way they respond to crustal shortening by increasing crustal thickness. Present-day mountain belts such as the Alps, the Himalayas and the North American Rockies all display thrusts that have undergone displacements of many tens or even hundreds of kilometres. A famous example of a much older thrust system is the **Moine thrust zone** of Northwest Scotland (Figure 6.4C), which has been known and studied since the 1880s.

Major thrusts occur in zones in which numerous individual thrusts are linked together in a complex manner, all contributing to the overall movement. In bedded rocks thrusts may describe a **staircase path** by alternately following a bedding plane then rising up a **ramp** to achieve a higher level (Figure 6.4A). Progressive movement of the upper thrust sheet causes these ramps to be transported across flat-lying strata to form **fold** structures (Figure 6.4B) in the overlying sheet. Continued movement of the thrust sheet may result in sticking of the original thrust plane and the development of subsidiary thrusts, which form in advance of the original ramp and result in a very complex assemblage (e.g. Figure 6.4C).

The marginal regions of mountain belts are typically marked by major thrust zones. These generally follow a pattern in which the outer

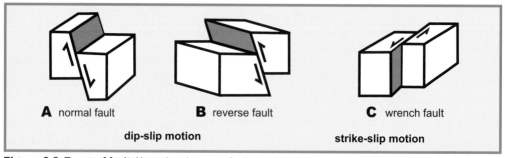

A normal fault   **B** reverse fault   **C** wrench fault

dip-slip motion   **strike-slip motion**

Figure 6.3 Types of fault. Normal and reverse faults involve dip-slip movement (up or down the fault plane); a wrench fault involves strike-slip movement (along the fault plane).

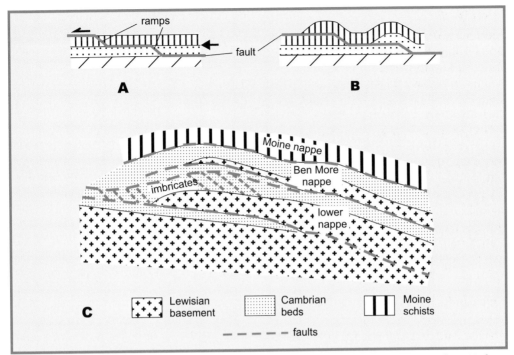

Figure 6.4 Thrust faults in profile. A, thrusts in layered rocks usually follow bedding planes for part of their course then cut up through the beds on a **ramp** structure. B, the result of this flat/ramp geometry is that the thrust sheet forms folds to accommodate to it. C, simplified diagrammatic W-E cross-section through part of the Moine thrust zone of NW Scotland, showing how successive thrust sheets (termed **nappes**) bring older, deeper rocks from further east upwards towards the west.

thrust sheets are formed from bedded rocks of the **foreland** (the region outside the mountain belt) whereas further inwards towards the interior of the belt they are replaced by much larger thrust sheets derived from deeper levels of the crust (as in Figure 6.4C). The latter are often quite different from the foreland rocks and include **metamorphic rocks** strongly altered by heat and high pressure, since the thrusts have reached down towards the base of the crust. It is because of this that we are able to study rocks formed at depths of up to 80 km beneath the surface.

Normal faults and crustal extension

Sets of normal faults (e.g. Figure 6.5) are formed in regions where the Earth's crust is being extended and thinned. Such regions occur at constructive plate boundaries – that is, along the ocean ridges and continental rift systems. Extensional fault systems also occur locally in mountain belts due to gravitational collapse of the thickened crust.

Normal faults are typically arranged such that uplifted and depressed blocks are bounded by sets of parallel or sub-parallel steeply-dipping faults, inclined away from

Figure 6.5 Normal fault: note that the thick bed has been displaced down to the left by the fault movement.

the uplifted blocks and towards the depressed blocks (Figure 6.6A). However, only limited extension can be achieved by this arrangement and further extension requires the faults to be rotated into a more gently inclined attitude. This has the result of also rotating the blocks, so that originally horizontal strata within the blocks also become inclined. In some cases, normal faults, as they descend into lower levels of the crust, may become curved (**listric faults**) or even sub-horizontal, allowing the fault blocks above them to rotate and slide over the basal fault which acts as a detachment horizon (Figure 6.6B). This allows much larger extensions to take place.

Strike-slip fault systems

Strike-slip faults (also known as **wrench faults**) are typically vertical or steeply inclined and separate blocks of crust that move past each other in opposite directions in a sub-horizontal manner (*see* Figure 6.3C). Sets of such faults are typical of conservative plate boundaries, of which perhaps the best-known example is the **San Andreas fault** in western USA (*see* Figure 5.7). This structure is actually a fault zone up to 100 km wide containing many individual faults that form a branching network. Because only a short section of a particular fault is active at any given time, the opposed movements of the blocks on either side create zones of compression and extension at the ends of the moving blocks. Subsidiary structures such as normal faults are formed in the extended zones and reverse faults in the compressed zones. Another complication arises because of bends in the main fault. Such bends swing either across or away from the main direction of movement creating zones of compression or extension respectively (Figure 6.7A, B). Active or recent strike-slip fault zones such as the San Andreas are marked by the close association of uplifted blocks that have been created in the compressional zones and depressed basins in the extensional zones.

Where strike-slip faults form part of the plate boundary network, they are known as **transform faults**. As mentioned in Chapter 5, the name 'transform' was applied to such faults because it was recognised that divergent motion at an ocean ridge, or convergent motion at a trench could be transformed into parallel motion along these faults, which could therefore be used to indicate the direction of relative motion of the plates on either side. Active transform faults in the oceans are very common and

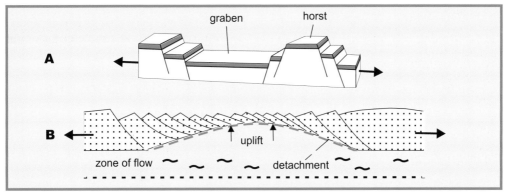

Figure 6.6 A. **Extension** on a set of steeply-dipping normal faults creates down-faulted and up-faulted blocks (termed respectively **graben** and **horsts**) but the red layer has not been much extended. B. Cross-section through the crust showing how larger amounts of extension cause initially planar faults to become curved and rotate due to solid-state flow in the lower crust. This process results in extension and thinning of the upper crust, leading ultimately to arching up of the lower-crustal material.

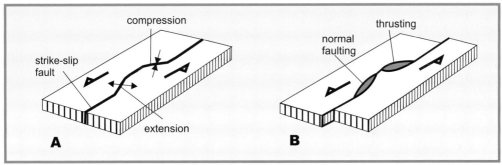

Figure 6.7 Bends in the course of a **strike-slip fault** create zones of **extension** and **compression** resulting in subsidiary normal faulting and thrusting, respectively. A, before movement; B, after movement.

6

are marked by periodic shallow earthquakes. Underwater surveying by side-scan sonar has identified linear ridges and troughs along the course of some of these faults.

Joint systems

All rock outcrops display numerous cracks or small fractures that generally lack obvious signs of displacement; that is, they are too small in scale to be classified as faults. Such cracks are termed **joints**. Joints typically form sets with similar orientations and origins. Because erosion preferentially attacks the joints, the rock outcrop may become carved into a series of blocks defined by the joint surfaces (e.g. *see* Figure 3.1).

The most common type of joint is extensional in origin, and formed either by contraction of the rock, or because it has been stretched. Such joints are often filled by vein material such as quartz or calcite that has been deposited by percolating solutions. Ore minerals may be hosted by such veins (*see* Chapters 1 and 10). In the ideal case, sets of extensional joints are perpendicular to the direction of extension of the rock but often the orientation of the joints is very variable.

Another type of joint is formed under compression and is similar to a small-scale wrench fault. Pairs of such joints may form in such a way that the smaller angle between them is bisected by the compression direction. This property is useful in identifying the orientation of the forces acting on the rock.

Other extensional joints are formed as a result of what is termed **unroofing**. This occurs when rock above the present ground level is stripped off by erosion, thus reducing the pressure on the rock below. The release of this pressure causes the rock to expand, typically forming sets of joints parallel to the bedding in horizontal stratified rocks, or sub-parallel to the present ground surface in igneous rocks, as shown by the example in Figure 3.1.

Contractional joints are associated particularly with igneous bodies where they have formed as the rock contracts on cooling. In some cases these joints form a set of polygonal columns at right angles to the cooling surface. The Giant's Causeway in County Antrim, Northern Ireland, and the island of Staffa, west of Mull in NW Scotland, are two well-known examples of this phenomenon. The columns of Staffa, formed within a lava flow (Figure 6.8), are vertical and have hexagonal cross-sections, like a honeycomb.

Figure 6.8 Columnar jointing in a horizontal basalt lava flow, Staffa, Inverness-shire, Scotland. Note the vertical joints forming a polygonal pattern and also the horizontal joints.

Fault rocks

Movement along a fault produces a zone of broken and crushed rock fragments of varying size. Such rocks formed near the surface may be composed of large angular fragments and are termed **fault breccia**; where the fragments are small, a type of clay is formed, termed **fault gouge**. Such rocks are soft and easily eroded, which explains the fact that faults are typically marked by valleys. Faults are also defined by lines of cliffs, where contrasting rock types, differing in their resistance to erosion, have been brought together by the fault movement.

At greater depth, fault rocks are much harder and may become partly recrystallised to form a special type of metamorphic rock. A fine-grained banded variety of such a rock is termed a **mylonite** and is found along major deep-level thrusts such as the Moine thrust of NW Scotland.

Squeezing and stretching – rock deformation

Rock layers may become bent when subjected to forces acting within the Earth's crust and can form quite complex shapes. Structures formed in this way are termed **folds**; some of the larger examples of these are among the most spectacular objects of the geological world. Folds can be formed in several different ways and assume a wide variety of shapes. Together with faults, they represent the response of the crust to the forces generated by plate tectonics and assume their greatest expression in mountain belts. The general term applied to the changes in the rocks brought about by these forces is **deformation**.

Fold shape

The simplest type of fold structure is produced by the upward or downward bending of a rock layer, and is termed respectively an **anticline** or a **syncline** (Figure 7.1A). Such a fold possesses two **limbs** joined by a **hinge**. The limbs of a fold may be curved, as in Figure 7.1A or straight, forming more angular folds – **chevron folds** (Figure 7.1B). The shape of a fold, or set of folds, is usually described by viewing the fold in profile, perpendicular to its hinge as in Figures 7.1 B–F. Fold sets may be described as **symmetric** if the opposite limbs are of equal length, or **asymmetric** if unequal (Figure 7.1C).

The shape of a folded layer is important in determining how the fold has been formed (the fold mechanism). Folds where the layers maintain their thickness through the fold are known as **parallel folds** (Figure 7.1D) and are formed by a process of **buckling** where a relatively strong layer has been subjected to a compressional force acting approximately parallel to the layer. Such folded layers are deformed internally such that the outer parts of the fold hinge are stretched and the inner parts compressed. In most folds however, the layers vary in thickness to a greater or lesser degree. In **similar folds** (Figure 7.1E), each layer has exactly the same profile; in such folds the layer thickens in the hinges and thins in the limbs in a regular manner.

Where a set of rock layers of varying strength are subjected to compression, the shapes of the resulting folds are determined by the way the strongest layers deform, which is usually by buckling, whereas the weaker layers accommodate to the shapes formed by the stronger layers (Figure 7.1F). Thus, for example, in a layered series of strong sandstone beds separated by weaker shales, the fold style may be dominated by parallel folds formed by the sandstones, whereas the weaker

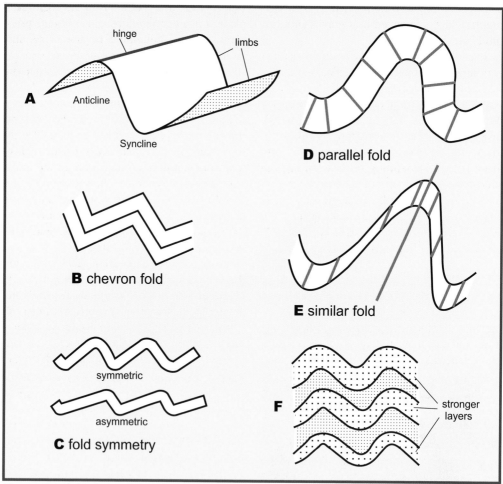

Figure 7.1 Fold shapes. A, anticline and syncline showing fold limbs and hinge. B–E: fold shapes in profile (perpendicular to the folded layer): B, chevron fold; C, symmetric and asymmetric folds; D, parallel fold; E, similar fold. In D and E the coloured lines within the fold layer are all the same length (in D perpendicular to the layer, in E parallel to the flow direction). F, set of folds showing buckling of stronger layers forming parallel folds whereas weaker layers form more similar folds that accommodate to the space between the buckled layers.

shales between may form folds that are more similar, or even quite irregular, in style.

Not all folds are formed by the active compression of a layer. Folds are also formed as a result of 'passive' bending in response to gravity, for example, as when a layer becomes draped over a normal fault, or to the displacement of a thrust sheet, as illustrated in Figure

6.4B. Folds may also be formed in soft sedimentary layers involved in a gravity-induced mud flow (*see* Chapter 4).

In contrast to faults, folds in solid rock form very slowly. Whereas a fault displacement may take only a few seconds, large folds may take hundreds or thousands of years to develop. An illustration of the rate at which this process takes place may be demonstrated by suspending a thin slab of sandstone between supports at each end. Depending on how strong the sandstone is, after months or years, the slab may be seen to have bent downwards under its own weight. Thicker slabs, such as the horizontal door lintels on old houses, are obviously unaffected after hundreds of years.

Mechanisms of folding

The way in which a rock layer deforms into a fold shape very much depends on both the material of which it is made (e.g. how strong it is) and the surrounding physical environment, especially the temperature and pressure.

Deformation at upper levels in the crust takes place at low temperatures and confining pressures and is dominated by fracturing and faulting. Displacements are achieved more by sliding between grains than by deformation within the grains themselves. Where rock layers are extended on the outer curves of parallel folds, for example, the extension is achieved by the opening of cracks, which are then filled by vein material such as quartz or calcite deposited from solution. Compression on the inner curves causes material to be squeezed out or dissolved, and transferred to the extending areas. Many folds are associated with faults, movement on which helps to determine the final shape of the fold (Figure 7.2A).

Deformation at deeper crustal levels, on the other hand, is controlled much more by the effects of higher temperatures and the greatly increased confining pressures caused by the weight of rock above. Under such conditions, deformation takes place by a process of flow in the solid state, which depends on the

7

Figure 7.2 Fold Mechanisms. A. This fold has formed in association with faulting (note displaced inner pale bed) indicating formation in the upper crust. B. These folds have formed partly by a process of flow at lower crustal levels than A. Note that the lower pink layer has accommodated to the shape of the grey layer above.

recrystallisation of the minerals making up the rock layers in such a way as to accommodate to the new shape of the deforming layer. Thus formerly round rock particles may recrystallise either as the same or different minerals with a more elongate and thinner shape. In many cases the individual layers deform passively, rather than actively as in the buckling process, and the flow direction may be oblique or even transverse to the layer being folded (Figure 7.2B).

Rock fabric

The deeper-crust deformation just described is accomplished by means of thoroughgoing changes to the rock that result in the development of a new structural 'texture', termed the **fabric**. This new fabric consists of changes to the shapes and orientations of grains, crystals and other objects making up the rock body in such a way as to reflect the change in shape of the whole rock body. Thus, for example, particles within the rock become flattened or elongate in the direction of greatest extension, and new minerals crystallise with their long axes in that direction.

If the undeformed rock body is represented by a cube, the deformed body becomes rectangular-sided, with the longest axis representing the direction of greatest extension and the shortest, the direction of greatest compression (Figure 7.3). The fabric, therefore, shows how the rock has responded to the forces acting on it. Layers parallel or sub-parallel to the short axis of the deformed body are therefore likely to be folded, and those oriented close to the long axis to be elongated. Elongated layers may be thinned (Figure 7.4A), or may form structures

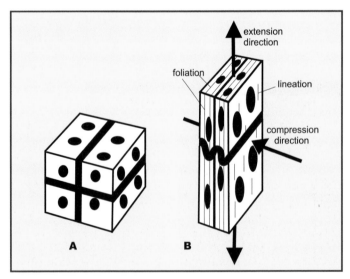

Figure 7.3 Rock fabric. A, undeformed block. B, deformed block showing directions of greatest extension and greatest compression. Note that the layer parallel to the compression direction has contracted and folded whereas the layer parallel to the extension direction has stretched. A planar fabric (foliation) is perpendicular to the compression direction and a linear fabric (lineation) is parallel to the extension direction.

called **boudins**; here the layer is pulled apart into sections, which have the appearance of a string of sausages (Figure 7.4B).

New planar structures formed in this way are termed **foliation**, and linear structures, **lineation** (Figure 7.4C). Fabrics may be wholly planar, or wholly linear, or have both planar and linear elements. Planar fabrics include the well-known **slaty cleavage** that enables roofing slates to be easily split; also **schistosity** and **gneissosity**. The latter two structures are found in coarsely crystalline metamorphic rocks (*see* Chapter 5) in which the nature of the fabric is obvious, whereas the grains in a slate are so small that their shapes and orientations can only be observed under the microscope. Schistosity consists of the parallel alignment of planar minerals such as mica, whereas gneissosity consists of alternating layers of light-coloured minerals – mainly quartz and feldspar, and dark-coloured minerals such as mica and hornblende.

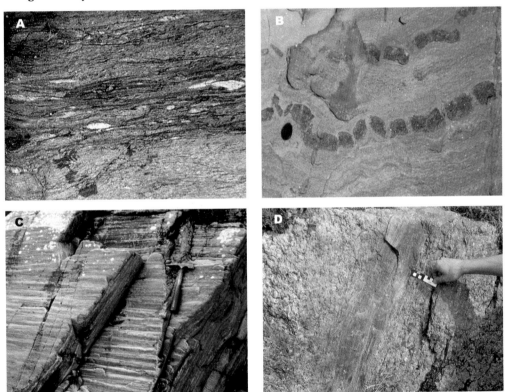

7

Figure 7.4 Fabric. A – foliation: deformed conglomerate showing pebbles flattened and stretched parallel to the foliation. **B – boudinage**: the dark layers have been stretched and separated into blocks. **C – lineation**: the linear structure on the surface of the shale beds is caused by the intersection of bedding and **slaty cleavage**. **D –** small **shear zone** in granite; a faint foliation can be seen on the left, curving into the shear zone.

Displacements in the deep crust: shear zones

The most intense deformation takes place in zones of displacement in the lower crust which correspond to faults in the upper crust and are known as **shear zones** (Figures 7.4D, 7.5).

Major thrusts, such as the Moine thrust illustrated in Figure 6.4C, as they reach down below about 15 km depth to warmer levels of the crust, become shear zones, where the displacement across them is taken up gradually, rather than abruptly as across a fault plane. Shear zones vary in thickness from centimetres to kilometres, depending on the amount of movement, type of rock and depth. The deformation within deep shear zones takes place by solid flow mechanisms as in deep-seated folding, but all gradations exist between these and mechanisms dominated by fracturing at higher levels.

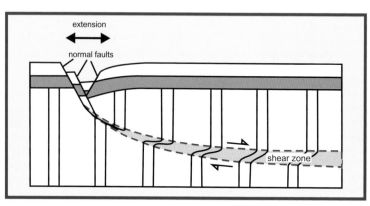

Figure 7.5 This diagrammatic section through the crust shows how displacement caused by normal faulting near the surface merges into a shear zone at depth.

Geological time and the age of the Earth

Relative dating

The principles underlying the relative dating of rocks arose out of the work of early observers such as Nicolaus Steno and James Hutton. Steno, in the 17th century, stated the principle of **superposition**, namely that younger sedimentary rocks were laid down horizontally above older sedimentary rocks. James Hutton, a century later, pointed out that the effects of processes that could be observed today, such as erosion, sedimentation, and volcanic activity, could be observed in the rocks. He emphasised the need for a vast geological timescale to accommodate the number, and slowness, of the events that could be read from the geological record. The general principles that arose from this early work, though not appreciated widely until much later, were as follows:

- Younger sedimentary strata lie above older (unless subsequently disturbed by folding or faulting);

- Igneous rocks that cut across, or cause metamorphic changes in, the surrounding rocks, must be younger than these rocks;

- Geological structures such as folds or faults are younger than the rocks that they affect;

- Rocks that contain components of other rocks (e.g. pebbles in sedimentary rocks or inclusions in igneous rocks) must be younger than the rocks from which the components are derived.

Hutton used the now famous 'Hutton's **unconformity**' at Siccar Point in Berwickshire, Scotland, to illustrate the principle of relative dating. At Siccar point, steeply-inclined beds (now known to be of Silurian age – see below) are cut across and overlain by gently inclined beds which are now known to be of Devonian age (Figure 8.1). Moreover the lowermost of the upper set of beds contained pebbles derived from the older set. Hutton inferred from this that several events had taken place: 1) deposition of the lower beds; 2) tilting of the lower beds to a steep attitude by the effects of severe earth movements; 3) erosion of the lower tilted beds so that the upturned beds are cut across by an old ground surface; 4) deposition of the upper set of beds using, in part, material derived from the lower set; 5) tilting of the upper set of beds to their present attitude; 6) erosion of the whole outcrop to its present shape.

These principles enabled a sequence of events to be constructed in any well-exposed

8

Figure 8.1 Hutton's unconformity: the upper set of beds (of Devonian age) lie above and cut across the eroded steeply-inclined lower set (of Silurian age). IPR/73–34C British Geological Survey. © NERC. All rights reserved.

area. However to be able to establish a geological timescale that could be applied over a whole country, or even worldwide, required a different technique – and William Smith, the English canal engineer, was one of the early pioneers in demonstrating that particular sedimentary strata could be identified by the fossils they contained, and recognised in different parts of the country. The use of fossils in the relative dating of rocks is discussed in the following chapter.

The stratigraphic column

Once a particular set of strata (e.g. a **formation** – *see* Chapter 4) could be identified from the fossils it contained, it became possible to establish a geological succession by following the formation across country until a different and younger formation could be identified lying above it, and another, older, formation lying below it. By this means, a succession was established of fossil-bearing strata commencing with the oldest, named the **Cambrian System**, up to the youngest, named the **Quaternary System**. Each of these systems contains a number of different formations. This geological succession is often termed the **stratigraphic column** (Table 8.1) and was eventually seen to apply over the whole world, with minor variations in the species of the particular fossils (called **index fossils**) used to identify the various

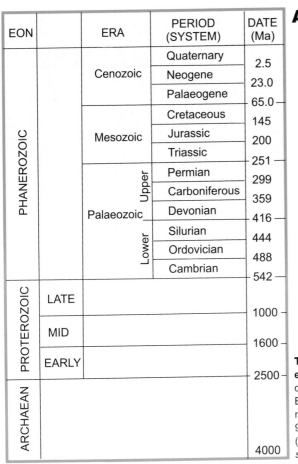

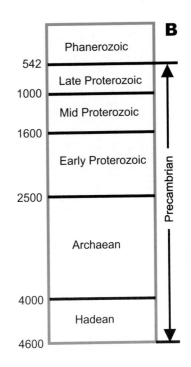

Table 8.1 A. The stratigraphic column – **eons**, **eras** and **periods** of geological time; dates given in million years before present. B. The stratgraphic column drawn to scale; note that the Precambrian represents nearly 9/10ths of geological time. (Periods are further subdivided into epochs – *see* Chapter 10.)

8

formations. Rocks older than the Cambrian were termed **Precambrian**. Not until much later was the Precambrian subdivided, nor any forms of life identified in it.

The rock units (**systems**) listed in the geological column also represent time periods, but it was not until the discovery of dating methods employing the decay of radioactive minerals (*i.e.* **radiometric methods** – *see* below) that precise times could be allocated to the various systems. So these various systems represent sets of strata and also periods of geological time. Before the ages of the different systems were established, geological time was subdivided using terms that refer to the forms of life in the different periods. Thus the whole of geological time since the Cambrian was called the **Phanerozoic Eon** (meaning age of visible life) and this was further subdivided into three **eras**: the **Palaeozoic Era** (age of old life), the **Mesozoic Era** (age of middle life) and the **Cenozoic Era** (age of young life). Thus we may refer to the **Cambrian System** (the rocks) and also to the **Cambrian Period** (the time). Once more was known about the Precambrian, it was in turn subdivided into the **Archaean Eon** (meaning ancient age) and the **Proterozoic Eon** (meaning age of former life).

Radiometric dating has shown that the oldest rocks so far discovered are around four billion (four thousand million) years old and that the age of the Earth is about 4.6 billion years old. Geologists usually employ the shorthand **Ma** for million years and **Ga** for billion years. Thus the age of the Earth is 4.6 Ga or 4600 Ma. It can be seen from the dates shown in Table 8.1 that the Precambrian occupies about 90% of geological time and the Archaean nearly half. It was during this almost unimaginably long period that early life appeared and

developed to the point when, at the beginning of the Cambrian, 542 Ma ago, an explosion of different life forms occurred, now recognisable because of their preserved hard parts.

It is difficult to visualise such long periods of time. One way to attempt this is to stretch out a piece of rope 4.5 m long to represent the age of the Earth. One millimetre of this length represents one million years. Even this length of time is difficult to imagine. One hundred years, which is easier to conceptualise, is represented by only 0.0001 mm of the rope's length!

Absolute dating: radiometric ages

The methods outlined above, involving the use of fossils and the principles of superposition, may give the impression, as they did to James Hutton, that the Earth is unimaginably old, but the calculation of actual ages of particular formations had to await the discovery of dating methods based on the radioactive decay of certain minerals. These techniques are termed **radiometric dating** and give what are called **absolute ages** as distinct from **relative ages**.

Radioactive elements change their atomic make-up through time at a uniform rate by emitting sub-atomic particles plus energy. The radioactive **parent** is said to decay and the products of this decay are known as **daughter** elements. The rate at which the new particles are formed is always proportional to the mass of the parent, and therefore declines at a steady and predictable rate through time. This means that if we measure the ratio of daughter to parent elements and know the rate of decay, we can calculate the amount of time in which the decay has taken place; in other words, the time since the original mineral was formed. A number of different radiometric dating systems are used based on commonly occurring radioactive elements, such as uranium, potassium, and rubidium. These methods are distinguished by their different half-lives. The **half-life** of a radioactive decay process is the time it takes for half the parent element to decay. For example, if we start with, say, 32 radioactive atoms, there would be 16 left after one half-life, 8 after two half-lives, 4 after three half-lives and so on. If the half-life of a particular process is 1000 years, say, and we find that one quarter of the parent element is left, the elapsed time since the beginning of the process must therefore be two half-lives, or 2000 years.

The half-lives of the various systems can be measured in the laboratory; these vary widely from seconds to thousands of Ma. Systems that have a relatively fast decay rate, such as carbon 14, can only be used for dating events a few thousand years old. On the other hand, Uranium 238, which decays to lead, has a half-life of 4500 Ma and can be used for dating very old Precambrian rocks.

The accuracy of these methods varies widely and most dates are given with error limits, in the form, say, 300 ± 10 Ma. Errors arise because of limits on the accuracy of the analytical process, and also because of natural variations in the rocks themselves. For example **whole-rock** dates lump together varying real dates of the various minerals in the rock whereas mineral dates tend to be more accurate. The most accurate dates are currently obtained from measuring the uranium–lead ratios in individual zircon crystals and commonly have stated errors of only a few million years.

The age of the Earth

James Hutton in the 18th century and Charles Darwin in the 19th, from different standpoints, both realised that the Earth had to be extremely old. Hutton saw the need for an enormous amount of time for the slow geological processes he envisaged to produce the rock sequences that were observed. Darwin, likewise, required this timescale for the process of evolution to produce the most complex forms of life from the simplest. Both, however, were opposed by the religious establishment of the time, which held that the biblical account of creation required the Earth to have been created in its present form in seven days. Indeed Archbishop Ussher in the 17th century, from studying the chronology given in the Old Testament, had concluded that the Earth was created in 4004 BC.

Early attempts to estimate the age of the Earth using scientific methods were made by Charles Lyell and others, from rates of sediment accumulation, giving ages of several million years. In 1897, Lord Kelvin gave an estimate of 20–40 million years by assuming that the Earth commenced as a molten body and cooled at a steady rate to its present state, and in the following century, John Joly made an estimate of 90 million years by measuring the amount of salt in the oceans.

However, the discovery of radioactivity around the beginning of the 20th century transformed the debate over the age of the Earth, and Arthur Holmes in 1913 produced the first geological timescale using radiometric methods. He also showed that Kelvin's calculation based on a cooling Earth seriously underestimated the age because the effects of the decay of radioactive minerals within the Earth provided a continuing supply of heat.

8

The currently accepted date for the age of the Earth is 4600 Ma. This date is based on the radioactive decay of uranium to lead. The starting point for the calculation is the uranium–lead ratio in present-day deep-sea sediments; this is thought to be an accurate representation of the present whole-Earth ratio because the deep oceans receive sediment from a wide variety of sources. Because some lead was present in the original Earth composition, this calculation gives an overestimate of the Earth's age. The original uranium–lead ratio when the Earth was formed is assumed to correspond to that found in meteorites, which are believed to have been formed at the same time as the Earth and other planets. This is used to calculate the original amount of lead, which is subtracted from the amount present in the sediments. This then gives an age of 4600 Ma for the Earth, which corresponds to the age calculated from the meteorites and also from the Moon, providing a check on the method.

8

Fossils – a history of life

Fossils are among the most interesting objects in the natural world. Finding a perfectly preserved specimen of an animal from its tomb within a rock is an exciting event and one that has prompted many young people to study geology.

Fossils have of course been known since ancient times and have occasioned much speculation as to their origins, but it was not until the early nineteenth century that the serious study of fossils (**palaeontology**) commenced with the work of Baron Cuvier in Paris. Cuvier studied the anatomy of animal skeletons and showed how fossil animals could be reconstructed from their incomplete fossilised remains. There was considerable controversy in those days about the significance of fossils. Some, the **catastrophists**, believed that fossils were the remains of animals that had perished during a single great event – the biblical flood; however, the view that prevailed held that fossils represented a catalogue of the animal and plant life that has existed over the long period of geological time. The latter view was strongly influenced by James Hutton's observations in the latter part of the 18th century that the Earth has a very long history, made up of a series of events similar to those that could be observed at the present time – erosion, sedimentation, vulcanicity etc., and that there was 'no vestige of a beginning, no prospect of an end'.

Nature and preservation of fossils

Fossils are the remains or traces of once-living organisms. Complete preservation of dead organisms is quite rare, although some remarkably well-preserved carcasses of the extinct woolly mammoth have been recovered from frozen tundra in Siberia, for example. In most cases the soft tissues of animals have been destroyed and only hard parts – like vertebrate skeletons and mollusc shells – are preserved. With the passage of time, the hard parts of fossils may be replaced by minerals such as calcite, silica or iron compounds, but the detailed structure is often preserved in minute detail so that the organism's anatomy can be reconstructed. In some cases the original material has been completely dissolved but the impression of the organism, called a **mould**, has been left imprinted in the rock. Even the most minute unicellular organisms

9

can be preserved, in silica for example, and can be studied under the microscope.

Tracks and other traces of organisms are also considered to be fossils – for example dinosaur footprints or worm burrows – and are valuable in enabling the palaeontologist to reconstruct the environment in which such animals lived, and something of their habits.

Only a tiny proportion of past life ever finds its way into the fossil record. There are various reasons for this. Much organic material is consumed by other life forms, or decays unrecognisably. The conditions for survival are untypical and depend on rapid burial, by mud or volcanic ash for example, at the bottom of the sea or in a lake. Consequently the fossil record has to be regarded as incomplete and unrepresentative. Many types of extinct organisms, especially those with no hard parts, will probably never be known.

Use of fossils in dating rocks

At about the same time as Cuvier in Paris was carrying out his studies of the nature and significance of fossils, William Smith, in his work as canal surveyor in England and Wales, was making detailed observations of how different fossils characterised different sedimentary strata. He found that a vertical order could be established, known as the **fossil succession**, which held true for widely separated parts of the country. Smith also observed that particular fossils, once they had disappeared from the succession, never re-appeared. Smith is often regarded as the founder of stratigraphy, although he did not realise the significance of the fossil succession in terms of the vastness of geological time and the concept of evolution – that insight was due mainly to Charles Darwin, as we shall see later.

Based on the work of Smith and others, a geological timescale was set up (as seen in Chapter 8), commencing with the Cambrian System (the oldest rocks then known to contain recognisable fossils) and ending with the Quaternary System, in which each unit recognised was distinguished by a particular assemblage of fossils. Rock sequences older than the Cambrian System were defined as Precambrian. Eventually these systems were subdivided into smaller units and also grouped together into larger units, but it was not until the advent of radiometric dating methods (*see* Chapter 8) that any precision could be applied to the dating system or any idea could be gained as to the real age of the various systems or of the Earth itself. Fossils offer a means of correlating rock successions from place to place in order to establish a time equivalence, both regionally and, ultimately, worldwide. They therefore provide a method of **relative dating** of rocks, that is they can be used to place a particular bed of rock in its correct position in the stratigraphic table as set out in Chapter 8, but the table itself must be calibrated by the **absolute dating** of various horizons within it by radiometric means.

Fossils are of no value in dating if they are found in strata of widely varying age; that is, if the organism has survived unchanged for long periods of geological time. Some fossils, such as the **brachiopod** *Lingula* for example (*see* below), are present in strata ranging in age from the Ordovician, 500 million years ago, until the present day. To be useful in dating, an organism must be: 1) limited in range to a short period of geological time – i.e. rapidly evolving; 2) widespread in occurrence; 3) found in a variety of sedimentary environments; and 4) have easily recognisable features.

9

Types of fossil and their classification

The system for classifying living organisms was introduced by the Swedish naturalist Linnaeus and subsequently extended to fossils by Cuvier and others. An organism is characterised as a **species** and given a **specific name**, if it can interbreed with, and produce viable offspring from, other members of that species.

Since fossils obviously cannot be observed interbreeding, the criteria for recognition of separate species must rely on differences in morphological features. Deciding which differences are significant is necessarily subjective, and has given rise to considerable debate. Palaeontologists tend to be either 'lumpers', who argue for a wide range of differences within a single species, or 'splitters', who favour finer distinctions between species.

Closely related species are grouped into a **genus** (plural, **genera**) so that an individual organism, such as the human being, is designated *Homo sapiens*, where *Homo* is the genus and *sapiens* the species. The genus *Homo* includes several other related species such as *Homo erectus*. Related genera are further grouped into **families**, families into **orders**, orders into **classes** and classes into **phyla** (singular, **phylum**), so that a hierarchy of relatedness is established (Table 9.1). Thus the genus *Homo* belongs to the family **Hominidae**, in the order **Primates**, in the class **Mammalia**, in the phylum **Chordata**. The Chordata, which includes all the vertebrates, is one of around a dozen phyla that together form the animal kingdom or **Animalia**. The plants form a separate kingdom.

Unicellular and simple multicellular organisms form two further groupings separate from the plants and animals, which are descended

KINGDOM	Animalia
PHYLUM	Chordates
CLASS	Mammalia
ORDER	Primates
FAMILY	Hominidae
GENUS	Homo
SPECIES	Homo sapiens

Table 9.1 The family tree of Man.

from them. These include **bacteria**, **algae** and **fungi**; various types of floating marine micro-organisms (**plankton**) which possess siliceous or calcareous skeletons are easily preserved. Simple organisms such as these are of enormous importance in the geological record. Not only are they the only form of life for the first 2000 million years, but as will be seen, they play an important part in transforming the environment of the early Earth.

Fossils and evolution

According to the theory of evolution, all forms of life are descended, via a multitude of intermediate stages, from a single ancestral organism, and their classification is based on the principle of relatedness. Closely related organisms share a recent common ancestor, whereas more distantly related ones have a common ancestor

9

further back in time. Closely related forms share the same basic body plan even though they may appear superficially very different (e.g. the elephant and the mouse are both mammals and are more closely related to each other than they are to the crocodile, which is a reptile). Thus mammals, reptiles, amphibians and fish are all vertebrates and possess certain features in common, such as a basic skeletal structure, but each of these groups originated at a particular point in geological time before which no representatives of their group are found in the fossil record.

Richard Dawkins, in his book *The Blind Watchmaker*, stresses that evolution should not be thought of as a progression from 'lower' forms of life to 'higher' as if man were standing on the top rung of a ladder on which successively lower rungs were occupied by, say, apes, reptiles, fish and so on. Instead, we should think of evolution as a bush with many branches, each diverging from its neighbours at different points but never converging. Thus all mammals are descended from a common mammalian ancestor which must have evolved (branched off) from a pre-existing vertebrate group (presumably a reptile) by means of some critical adaptation which conferred a significant advantage over its reptilian colleagues.

The concept of evolution is supported by the observation that relatively simple life forms characterised the older geological systems and that progressively more complex organisms appeared later in the geological record. While many simple types of organism survive and are successful to the present day, the converse is not true, that is, there are no complex types represented at all in the first 2000 Ma of Earth history. Thus it seems that simple forms give rise to complex forms over time. The process

of evolution can be demonstrated by following the gradual steps by which certain groups of animals change through successively younger strata. Good examples of this include the **graptolites** and the **cephalopods**, in both of which small changes in form occur with time that are interpreted as improvements in adapting to their environment (Figures 9.1, 9.2).

Such observations by themselves do not 'prove' evolution. The incompleteness of the fossil record is such that in some cases we may never see the intermediate forms ('missing links') through which one species gave rise to another. However, that the process can occur is proved by the selective breeding of domestic animals, whereby very different forms (of horses, dogs, etc.) can arise from a single ancestor.

In 1859, Charles Darwin in his book 'Origin of Species by means of Natural Selection' first demonstrated how competition in the natural world causes forms that are better adapted to their environment to compete more successfully with others (for food, or space, or to avoid predators) and ultimately to replace them. Thus a succession of tiny changes produced by genetic mutations, if they were advantageous in enabling the organism to compete, build up over time to produce differences so marked as to distinguish the organism as a separate species. An essential step in the process of creating a new species is the physical separation of individuals possessing a genetic advantage and thus preventing them from interbreeding with the rest of the population. If, for example, a population is divided by a physical barrier, such as a body of water or a mountain range, the two halves can evolve separately into different species that are no longer capable of interbreeding.

9

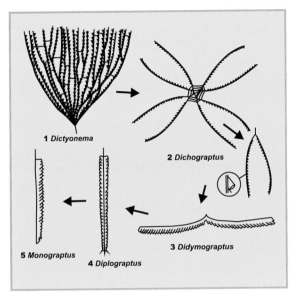

Figure 9.1 Evolution of the graptolites.
The **graptolites** are colonial floating organisms thought to be descended from complex branching forms such as *Dictyonema* (1) (Cambrian–Carboniferous) which were anchored to the sea bed or to floating objects. Their evolution through time involved a reduction in the number of branches from eight in *Dichograptus* (2) (early Ordovician) to two in *Didymograptus* (3) (early Ordovician) to two joined back-to-back in *Diplograptus* (4) (later Ordovician) and one in *Monograptus* (5) (Silurian). There was also a change from forms that hung downwards so that the individual organisms (**polyps**) in their cups (**thecae**) faced down as in 2 and 3 to the later forms such as 4 and 5 where the cups faced upwards. Changes also took place in the shape of the thecae. Evolution tended to produce greater simplicity of structure that may have made the colony more efficient.

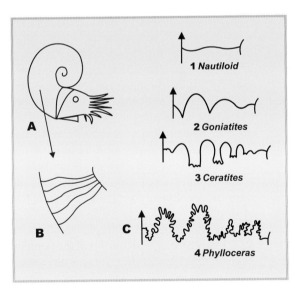

Figure 9.2 Evolution of the cephalopods.
Cephalopods are complex free-swimming **molluscs** with large brains, eyes and a set of tentacles for seizing prey (A). Their shells are cylindrical tubes divided into a series of chambers by partitions, marking the successive growth stages of the animal (B). These chambers were used as buoyancy tanks to enable the animals to adjust to some extent to varying water depths by pumping water in or out of the chambers. Their evolution (C) took the form of a gradual increase in the complexity of the **sutures**, which mark the intersections of the partitions with the shell. Note the changes in suture form from (1) *Nautiloid* (Cambrian-Recent) to (2) *Goniatites* (Carboniferous) to (3) *Ceratites* (Triassic) to (4) *Phylloceras* (Jurassic–Cretaceous). Arrows point to shell aperture. Increasing complexity conferred greater strength on the chambers, enabling the animals to withstand the pressures of greater depths of water.

9

Critics of evolution theory have raised the problem of creating very complex structures, such as the mammalian eye for example, by means of successive small changes brought about by favourable mutations. However, Dawkins points out that each tiny change to a light-sensitive structure would confer an evolutionary advantage, and that, given the enormous number of generations involved over periods of tens of millions of years, the natural process of random mutation would generate sufficient favourable changes for such structures to emerge.

Some important fossil groups

A number of fossil groups are important either because they are frequently preserved, or because they are useful in dating rock strata, or because they are exciting to the collector. The following selection ignores other groups that may have been more important in terms of numbers but have either not been preserved to the same extent, or are not stratigraphically 'useful'.

● *Early life.* The first forms of life on Earth are thought to have been **bacteria** deriving their energy from volcanic vents on the sea floor. Bacteria are single-celled organisms without a nucleus and represent the most primitive form of life. Some are **photosynthetic** – that is, they employ sunlight as a source of energy and release oxygen. The oldest structures identifiable as organic are **stromatolites,** which are found in 3.4–3.5 billion-year-old Archaean rocks and appear to record the only form of life during the Archaean. They are thought to be formed by colonies of bacteria, which lay down layers of calcite forming a series of mounds on the sea floor.

During the Proterozoic, **blue-green algae** (**cyanobacteria**) appeared and contributed to the stromatolite development during this period, when stromatolites were much commoner. Stromatolites still exist at the present day, although they have been much reduced in importance since the Cambrian Period due to competition by other organisms. The blue-green algae were eventually to give rise to the first plants (*see* below). Blue-green algae are photosynthetic, and their activity transformed the environment of the early Earth from oxygen-poor during the Archaean to oxygen-rich during the Proterozoic, thus enabling higher life forms to develop.

Towards the end of the Precambrian, more complex single-celled animals evolved, called **Protozoa**. These fed on other organisms, including algae. Two important **planktonic** (floating) types are **Foraminifera** and **Radiolaria**. The foraminifera had calcareous exoskeletons and were abundant during the Mesozoic where they have been widely used by oil companies in dating strata. The radiolaria had siliceous skeletons and form siliceous layers (**radiolarian chert**) on the deep ocean floor where no other sedimentary or organic matter survives. These planktonic organisms existed from mid-Cambrian times to the present day, and are important in dating certain oceanic strata that would otherwise be impossible to date.

● *Graptolites.* These belong to an extinct class, which is thought to belong to the phylum **Hemichordata**. The graptolites were colonial organisms in which a number of branches were joined to a single central body, rather like some seaweeds (Figure 9.1). Some floated while others were attached to the sea

9

Figure 9.3 A, a **goniatite** (a sub-class of the Cephalopoda) polished to show internal structure: note that the chambers have been infilled with crystalline material. Courtesy of the Orcadian Stone Company Museum, Golspie, Sutherland.

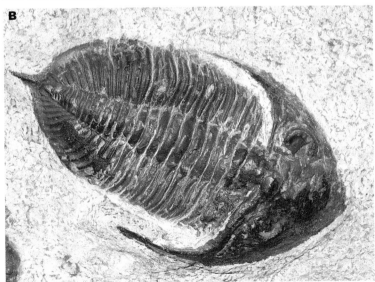

B, **trilobite**: note the three-lobed structure and armoured head-shield with large eyes; the body segments can move independently whereas the tail segments are fused together. Courtesy of the Orcadian Stone Company Museum, Golspie, Sutherland.

9

floor. Each branch contained an array of tiny cups containing the individual organisms, which fed by filtering sea water. The individuals were connected by a canal to others along the branch to enable them to share food. Generally only the hard skeletons of the branches survive, lying on the mud of the sea floor. Graptolites were abundant during the Ordovician and Silurian periods; the rapid changes in their skeletal arrangements through a succession of different forms has meant that they have become important indicator fossils in the subdivision of these periods. They became extinct during the Carboniferous.

● *Corals.* These, along with the sea anemones and jellyfish, belong to the phylum **Coelenterata**. The coelenterates are jelly-like marine organisms consisting of a single body cavity with a mouth surrounded by tentacles that wave in the sea water to catch passing food particles. Corals (essentially sea anemones with skeletons) range from single cup shapes to colonies of complex branching forms. They are basically conical or cylindrical, with radially arranged calcareous platelets giving strength to the structure (Figure 9.4A). The skeletons are an important component of many limestones, especially in the Silurian to Devonian periods; they have been used as index fossils in the Lower Carboniferous, owing to the rapid evolution of their skeletal form. Reef-building types of coral survive to the present day and are restricted to warm shallow water. They are considered to be an important palaeoclimatic indicator in the geological past (e.g. *see* Figure 5.2).

● *Brachiopods.* The phylum Brachiopoda (lamp shells) consists of invertebrate marine animals, more complex than the corals or graptolites, which were abundant in the shallow seas of the Palaeozoic Era and are still represented at the present day. They are single organisms inhabiting two calcareous shells (Figure 9.4B), an upper and a lower, which opened to allow the animal to feed by filtering the sea water. Most were attached to the sea floor by means of a short fleshy stalk, but some lay half-submerged in the mud. The complex arrangement of muscle structures to accomplish the opening and closing of the shells, together with the often detailed ornamentation of the exterior of the shells (Figure 9.4B), means that they are comparatively easy to identify, thus enabling the different types to be distinguished. Because some types preferred shallow water and others deeper, these fossils are useful in characterising changes in water depth reflected in the sedimentary strata, particularly in the Lower Palaeozoic, where they were abundant.

● *Echinoderms.* The phylum **Echinodermata** includes a wide variety of marine fossil types and is represented at the present day by animals such as the sea urchins and starfish. The organisms have calcite skeletons and a characteristic five-sided symmetry (Figure 9.4C). Their bodies are enclosed in a mosaic of regularly-shaped plates through which project tube-like appendages; these could be used for catching food, moving about or, as in the case of the sea urchin, modified into spines for defence. They occupied a variety of environments and life styles: some were static filter feeders, some moved along the sea floor grazing and others were predators. Two groups are particularly important in the geological record: the crinoids and the echinoids. In the **crinoids** (sea lilies) the body is mounted on a long stalk made up of ring-shaped segments.

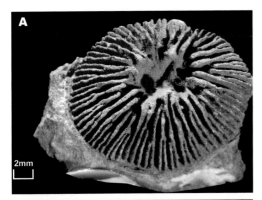

Figure 9.4 A, Ordovician coral: note radial partitions. B, fossil brachiopod *Terebratulina* from the late Cretaceous Chalk formation (~8cm across); C, Recent echinoid, *Clypeaster* – note that the basic five-fold symmetry is modified to give a more directional bilateral symmetry. D, ammonite, showing ribbed ornament. A, B, C, British Geological Survey. ©NERC. All rights reserved. IPR/73-34C, 122-06CT.

The stalk is attached either to the sea floor or to floating objects. A set of long arms extend upwards from the body to catch food. Whole crinoids are rarely preserved but the distinctive calcite segments are easily recognisable and form a significant component of many limestones of the Silurian to Carboniferous systems. They are still represented today. The **echinoids**, which include the present-day sea urchins and starfish, are represented from Ordovician times. Forms such as *Micraster* are common in the Chalk formation of Northern Europe.

● *Molluscs.* The varied phylum **Mollusca** contains the important class **Cephalopoda** (e.g. Figure 9.3A) together with the **Gastropoda** (snails, whelks etc.) and **Bivalvia** (mussels etc.). All marine molluscs have a

9

calcareous shell consisting of a hollow cone with numerous growth stages represented by successive rings (e.g. Figure 9.3A). The organisms have a mouth, a gut and paired gills. In the land-dwelling gastropods (e.g. the snails) the gills have been replaced by air-breathing lungs. The **rudists** are a group of reef-building bivalves which became like corals in their form and way of life. The **cephalopods** possess a head with eyes, and are free-swimming predators (Figure 9.2A). The best-known and most stratigraphically useful are the **ammonoids** (Figure 9.4D), in which the typically coiled shell is internally divided into chambers by calcareous partitions. The animal inhabits the outermost of these chambers, using the others as buoyancy tanks to assist it in swimming at varying depths. Ammonoids first appeared in the Devonian Period and became extinct at the end of the Cretaceous. Because of the rapidly increasing intricacy of the shape of the partitions, which trace complex suture shapes on the shells (Figure 9.2C), these animals are very useful as stratigraphic indicator fossils, especially during the Triassic and Jurassic periods.

● *Trilobites.* These are an extinct class of marine organisms belonging to the phylum **Arthropoda**. They have an armoured external skeleton and, as the name suggests, possess a central ridge and two side lobes (Figure 9.3B). They are divided into a head shield, a body and a tail, the latter two of which are sub-divided into many segments, to each of which is attached a pair of limbs used for walking, swimming or handling food. They were abundant in the shallow seas of the continental shelf during the Cambrian and Ordovician Periods; some were free-swimming while others were bottom-dwelling. Trilobites

probably evolved from soft-bodied worm-like ancestors in the late Precambrian and became extinct during the Permian.

The Arthropoda also include insects, spiders and crabs. The **insects** alone account for more species than the rest of the animal world together, but are poorly represented in the fossil record (Figure 9.5D).

● *Vertebrates.* These are of interest to us, both because they include so many familiar animal types and also because they show us an evolutionary pathway from the most primitive early forms over 500 million years ago to ourselves. Vertebrates are actually a sub-phylum of the phylum **Chordata**; members of this phylum are characterised by having a brain and a spinal cord running through the length of the body, giving the animals the ability to perform more sophisticated movements than any of the preceding types. Vertebrates protect this spinal cord by means of a bony internal skeleton.

The earliest vertebrates were the **fish**, which were cold-blooded, lived in water, and breathed oxygen through gills. Early fish were jawless and possessed body armour of bony plates to protect them from predators; these were common in the freshwater lakes of the Silurian and Devonian periods. The first jawed fish, which had scales and spines, appeared about 400 million years ago. Bony fish, with complete skeletons and fins (Figure 9.5A), also date from the Devonian and of course are abundant at the present day; one type developed lungs in the late Devonian and was able to survive on land.

The **amphibians** appear to have evolved from a fish-like ancestor around 350 million years ago. They were the first vertebrates to have limbs, which were used to move about

9

Figure 9.5 A, Jurassic fish, *Dapedium*: note the teeth, the long upper and lower fins, and that the scales, unlike modern fish, do not overlap. Courtesy of the Orcadian Stone Company Museum, Golspie, Sutherland. B, head of *Tyrannosaurus rex* skeleton. E446/0521 ©M.H. Sharp/Science Photo Library. C, cast of fossil remains of Archaeopteryx, a flying reptile with feathers, found in Jurassic fine-grained limestones at Solenhofen, Bavaria. Archaeopteryx had many of the features of a small dinosaur with the addition of feathers and had bird-like claws on its wings. It was crow-sized. E445/0118 ©Jim Amos/Science Photo Library. D, fly and spider trapped in amber. British Geological Survey. ©NERC. All rights reserved. IPR/122-06CT.

9

on land. They still required water, however, in which to lay their eggs and in which their young developed. Early amphibians dominated the swamps of the late Carboniferous and early Permian periods hunting for fish and insects.

Reptiles first appeared around 300 million years ago, having evolved from an amphibian ancestor by developing the technique of fertilising their eggs inside the body of the female. They were thus no longer partially tied to water, as were the amphibians, and colonised higher, drier lands between the river valleys of the Permian and Triassic periods. Like the amphibians, they were cold-blooded. In addition to the modern crocodiles, snakes and lizards, they include the very successful and well-known **dinosaurs** (Figure 9.5B). There cannot be many people who have not heard of, nor seen pictures of, huge animals such as *Tyrannosaurus* or *Stegosaurus* or watched a *Velociraptor* running across the television screen! In addition to these land-dwelling types, the dinosaurs also colonised the seas (**ichthyosaurs**) and the air (**pterosaurs**). Some were vegetarians while others were carnivores. They became extinct at the end of the Cretaceous Period.

The **birds** differ from the reptiles in three main ways: they are warm-blooded, have a covering of feathers, and their front limbs have been modified as wings capable of gliding or flying. Early birds such as *Archaeopteryx* (Figure 9.5C) had teeth and a bony tail; they were also probably too heavy to be capable of proper flight and may have glided from trees. Later types have toothless beaks and light-weight tails with feathers; their bones are air-filled to provide buoyancy. The first birds evolved about 140 million years ago from a reptilian ancestor.

Mammals seem also to have evolved from a reptilian ancestor about 190 million years ago. They differ from the reptiles in being warm-blooded, having a covering of hair and a more efficient heart and lungs. The **placental mammals**, instead of laying eggs, nurture their young in the womb and give birth to live babies, whom they feed with milk. Early mammals still laid eggs however, and an important group of mammals, the **marsupials**, nurture their young in a special pouch. The mammalian ancestors, which appeared in the mid-Cretaceous, were small mouse-like creatures that co-existed with the dinosaurs, but when the latter became extinct about 65 million years ago, these mammals evolved into a wide variety of types inhabiting many different environments including the air and the sea.

● *Plants.* The simplest plants are the algae, which have been abundant throughout geological time since the early Precambrian. They form the basic food supply at the bottom of the food chain on which the higher forms of life depend, directly or indirectly. Only those that had calcareous or siliceous coverings have survived as fossils. These include **coccoliths** and **diatoms**. Coccoliths had a calcareous covering and are an important component of Mesozoic marine limestones. They are present today as **plankton** (floating micro-organisms) in the upper levels of the oceans. **Diatoms** are similar micro-organisms but possess a siliceous covering. They are preserved as fossils in fine deep-water muds formed at such great depths that any calcareous material would have been dissolved.

More complex plants evolved from algae by developing a system of internal channels (a

9

vascular system) protected by cellulose walls whereby water and nutrients can be transferred to all parts of the organism. Early forms of vascular plants include both spore-bearing and seed-bearing types: the club mosses, conifers and ferns. They appeared about 400 million years ago, and achieved great importance in the Carboniferous System, where a variety of plants colonised the coastal swamps and gave rise eventually, after deposition and compaction, to much of the world's coal deposits. The great tree-like **club mosses** of the Carboniferous achieved heights of up to 30 metres. The more recent flowering plants probably appeared around 120 Ma, and by the beginning of the Cenozoic Era were the dominant type of plant.

Fossils and plate tectonics

Fossils are useful in plate tectonic reconstructions in two main ways: they can identify past environments (e.g. continental, marine etc.) and they can track former geographical locations. Different environments are represented by different fossil types. For example, land animals, freshwater bivalves and many plants indicate a continental environment; other plants indicate coastal swamps. Certain bivalves and brachiopods inhabit near-shore waters while others are representative of the deeper water of the continental shelf. Corals indicate clear, warm and relatively shallow marine conditions. The deepest-water conditions, far from any sediment supply, are represented by the radiolarian cherts, and so on. So from evidence such as this, one can interpret the history of various pieces of crust and reconstruct the shapes of former continents.

In **orogenic belts** (*see* Chapter 5), pieces of crust representing widely different environments are often brought together. For example, if deep-ocean radiolarian cherts have been placed next to continental rocks, it is an indication of the presence of a **subduction zone**.

Certain fossils can also be used to reconstruct the movements of former continents because they are geographically restricted to particular areas, while in other areas the same environments are represented by different species. Thus the Cambrian and Ordovician shallow-water marine fauna of NW Scotland is represented by species of brachiopod and trilobite different from those of England, but identical to those of eastern North America. This, together with other evidence, indicates that Scotland was attached to North America until the beginning of the Caledonian orogeny in the Silurian Period when a separate microcontinent, of which England and Wales were a part, collided with it (*see* Chapter 11). When the North Atlantic Ocean was formed in the Eocene Period, the whole of the British Isles was separated from North America. A collection of fossils within a specific area, which shows a uniform range of species but differs from that in other areas, is termed a **faunal province**.

Another indication of the value of fossils in plate tectonic reconstruction is provided by the breakup of the supercontinent Gondwana (*see* Figure 5.1). The land animals belonging to the Permian and Triassic systems throughout Gondwana belong to a single faunal province, whereas, after the break-up of that supercontinent in the Jurassic Period, the various faunas of each of the modern continents evolved in different directions, so that the large land animals of Africa, for example,

9

which are placental mammals, are completely different from those of Australia, which are marsupials. The time when the fauna and flora of two adjacent continents begins to diverge is an indication of the time of their break-up.

Mass extinctions

Fossil evolution seems not to have been a slow gradual process over geological time but is characterised by relatively sudden bursts of activity and by other periods when huge numbers of species become extinct. At certain times, between 50 and 90% of all species on Earth vanished over geologically quite short intervals. Such **mass extinctions** are documented in the fossil record at the end of the Ordovician Period around 440 Ma ago, during the late Devonian (around 365m Ma), at the end of the Permian, and, the widely known one, at the end of the Cretaceous (65 Ma), when the dinosaurs became extinct. All these mass extinctions were followed by a burst of activity when a relatively small number of species took advantage of the lack of competition. But this was followed by a period when new species evolved that were better adapted to the changed environmental conditions.

There has been much speculation as to the cause of the end-Cretaceous mass extinction, the most popular theory attributing it to a giant meteorite impact. Although this almost certainly occurred, it is unlikely to have been the sole source of the extinction, which in many species took place gradually over millions of years. There are many types of process that can affect the environment, including large changes in sea level and unusually severe volcanic activity. One of the most important is likely to have been periods of widespread glaciation that will have displaced many species to warmer parts of the globe and extinguished others. It has been suggested that at certain times glacial conditions may have extended over the whole Earth (the '**snowball Earth**'). This would have had a dramatic effect on Earth's population of organisms.

9

Turning the pages – Earth history

As explained in Chapter 8, geological time is divided into four first-order units, termed **eons**: the Hadean (4.6–4.0 Ga),the Archaean (4.0–2.5 Ga), the Proterozoic (2.5 Ga–542 Ma) and the Phanerozoic (542–0 Ma). The short-hand forms of Ga = thousand million years and Ma = million years are used throughout this chapter. The Phanerozoic is further subdivided into second-order units named **eras**: the Palaeozoic, the Mesozoic and the Cenozoic, and each of these in turn into third-order units named **periods**. In this account, for convenience, the Proterozoic is divided into Early, Middle and Late sections and the Palaeozoic into Lower (early) and Upper (late). These are not formal time units, but are commonly used by geologists as convenient subdivisions.

The first 600 Ma: the Hadean Eon

There is no direct evidence concerning the earliest part of Earth history, from 4.6 to around 4.0 Ga, since no rocks of this age have been found. Apart from the 4.6 Ga meteorites mentioned in Chapter 8 (which are much more recent arrivals) the oldest dated material consists of zircon crystals with ages of around 4.0 Ga, indicating that solid crust with a broadly granitic composition existed at that time.

Earth's earlier history is therefore necessarily speculative and depends upon assumptions about the likely way the solar system evolved. The planets are generally thought to have formed by the amalgamation of solid particles which orbited around the sun during the early stages of development of the solar system, gradually coalescing into rings and then into individual planets, plus numerous smaller bodies – the asteroids and the meteorites. The accretion process would produce heat arising from the gravitational energy of the colliding particles. The fact that the Earth is segregated internally into heavy elements (e.g. iron) in the core and lighter elements (e.g. silicon) in the crust, indicates that the Earth must have been heated up sufficiently to become molten not long after formation. The same process would have been responsible for the formation of the early atmosphere, which is thought to have corresponded to the mixture of gases presently emitted by volcanoes, principally water vapour, hydrogen, carbon monoxide, carbon dioxide and nitrogen. Free oxygen did not appear until later.

10

An important feature of these early stages would have been a series of collisions with meteorites and asteroids. The Moon's surface is covered by very numerous impact craters, which have survived because there has been no subsequent tectonic disturbance. It follows that large numbers of these bodies were passing through the solar system during this early period and must have affected Earth to the same extent. It has been suggested that the Moon itself may have resulted from a collision, or near-collision, between Earth and another planetary body.

At some point during this early period the crust and mantle must have solidified and the water vapour in the atmosphere condensed to become surface water. The original heat supply deriving from the Earth's formation would gradually be replaced by heat from the decay of radioactive elements which would have been preferentially concentrated in the upper layers of the Earth. Because this heat supply is limited, and radioactive decay proceeds at a steady rate, this heat source must gradually decline at a predictable rate through time, and the rate of heat loss through the surface during the early Precambrian would have been much greater than it is at the present day.

The Archaean Eon (4.0–2.5 Ga)

The earliest well-preserved rock sequence occurs at Isua in southern Greenland and pre-dates a metamorphic event at around 3.8 Ga. This sequence contains water-lain sediments and volcanic rocks very similar to those formed at the present day and confirms that the surface of the Earth was cool enough to support surface water at that time. Because of the higher heat flow, however, it is probable that volcanism was much more widespread and, rather than the large stable oceanic plates of today, the surface then may have been dominated by **hotspots** (*see* Chapter 5) rather than by a regular set of ocean ridges. The early crust would have been made up predominantly of basic volcanic rocks like the present-day oceanic crust, whereas the largely granitic crust of today would have evolved gradually through time, forming ever larger continental masses.

Most Archaean sediments appear to be of relatively deep-water origin, and include **greywacke** (*see* Chapter 4), mudstones and iron-rich **cherts**. The volcanic rocks include a high proportion of **ultrabasic lavas** (*see* Chapter 2) that are only produced in regions of unusually high heat flow – thought to be the equivalent of the present-day **mantle plumes** (*see* Figure 5.11). The iron-rich cherts consist of alternating thin layers of iron-rich material and silica and are termed **banded iron formations**. The chert is believed to have formed by chemical precipitation on the sea floor close to volcanic sources, and it has been suggested that the iron oxides may have been deposited by algae. Shallow-water sediments such as limestones and sandstones are relatively uncommon, suggesting the absence of large areas of continental shelf. Much of the record of Archaean conditions is preserved in the so-called **greenstone belts,** which consist of sequences of sediments and volcanic lavas surrounded by granite intrusions. These **granite–greenstone terrains** are found within the Archaean cores of all the continents.

All Archaean terrains contain abundant granitic rocks, which indicates that this was the time when large areas of continental crust were formed. These became the nuclei of the present continents, but it was probably not until late in Archaean times that any large continental masses developed. The earliest stable piece of continental crust has been identified in Southern Africa, where an area of about 40,000 km² is covered by a thick layer of non-marine sediments, including the famous Witwatersrand gold deposits, dated at 2.8–2.5 Ga. By the end of the Archaean, at 2.5 Ga, all the pieces of continental crust previously accumulated are thought to have come together to form a single large continent, or **supercontinent**.

Simple cellular structures thought to represent bacteria are the only form of life recorded in Archaean rocks. The oldest, probably around 3.5 Ga old, have been found in Western Australia and are associated with **stromatolites** (*see* Chapter 9).

The Early Proterozoic (2.5–1.6 Ga)

During this period, for the first time, there is evidence of the operation of plate tectonic processes in something like their present form. Large stable continental masses (called **cratons**) are flanked by **orogenic belts** and are cut through by swarms of igneous dykes that indicate that the cratons have been subject to extensional forces in the same manner as today. There is **palaeomagnetic** evidence (*see* Chapter 5) that, at the end of this period, Archaean cratons in North America, Greenland, Siberia and Scandinavia had collided to form a large **supercontinent** with collisional orogenic belts marking the positions of the joins between them (Figure 10.1). These orogenic belts contain the igneous and sedimentary products of subduction and subsequent collision over a period between about

1.9 Ga and 1.4 Ga. They include the **Hudsonian orogeny** of Canada, the **Nagssugtoqidian** of Greenland and the **Lapland–Kola** of northern Scandinavia, and are represented in NW Scotland by the **Laxfordian orogeny**.

Evidence of stable cratons of this age is present in every continent, and in many cases these cratons are partly covered by shallow-water deposits. Stromatolites are abundant, indicating that early life in the form of bacteria and blue-green algae was flourishing. The algae by this time are thought to obtain their energy from sunlight and to release oxygen, which eventually would provide the first oxygenated atmosphere. This process was likely to have taken hundreds of millions of years.

Banded iron formations are also very common but are confined to the early part of the Proterozoic, before about 2.0 Ga. They account for most of the present-day commercial iron ore deposits. The occurrence of these banded iron deposits indicates that free oxygen was available in the oceans, probably released by algae, as suggested above. At the beginning of the Proterozoic, continental deposits show no evidence of free oxygen, indicating that oxygen was not yet present in the atmosphere, but by about 2.2 Ga, continental red beds appear for the first time. These red beds require oxygen to be available to combine with iron to form the mineral hematite, and provide evidence that by this time oxygen was being released into the atmosphere.

Some of these early Proterozoic sequences also contain the first evidence of a widespread glaciation. The **Gowganda formation** of northern Ontario in Canada (*see* Figure 4.4A) contains boulders with glacial scratch marks and glacial clays with pebbles dropped by floating ice.

10

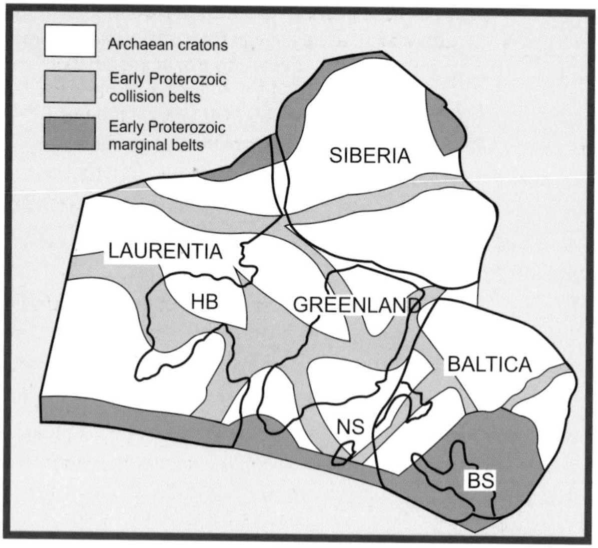

Figure 10.1 Reconstruction of part of an early Proterozoic supercontinent, showing how several stable Archaean blocks (**cratons**) have come together along early Proterozoic collision belts. Marginal belts where subduction is continuing are shown along two sides of the supercontinent. Both North America and Baltica are rotated with respect to their present orientation: note the positions of Hudson Bay (HB), the Baltic Sea (BS) and northern Scotland (NS). Continental outlines are different from today's since younger crust has subsequently been added to them. Based on Buchan *et al.* (2000).

The Middle Proterozoic (1.6–1.0 Ga)

During this period, more evidence is available supporting the operation of modern plate-tectonic processes. By fitting together the now-separated fragments of 1.0 Ga orogenic belts in all the present-day continents, it is possible to show that, by the end of the middle Proterozoic, a large number of separate continental pieces had amalgamated to form a single super-continent, named **Rodinia** (Figure 10.2). The central core of this new supercontinent was the continental mass formed at the end of the early Proterozoic by the amalgamation of **Laurentia** (North America plus Greenland), **Baltica** (Scandinavia plus parts of northern Russia) and Siberia. This reconstruction places the present-day continents in unusual positions: Antarctica adjoins western North America, and Africa and South America adjoin eastern North America. The best-studied of the collisional orogenic belts formed at this time is the **Grenville belt** of

eastern USA and Canada, which is thought to have resulted from collision with parts of Africa and South America containing fragments of a 1.0 Ga orogenic belt.

The Late Proterozoic (1000–542 Ma)

This period saw the transition from early uni-cellular life in the form of bacteria and algae to the extraordinary explosion of varied and relatively complex life forms that took place between about 700 Ma and 600 Ma, before the beginning of the Palaeozoic Era. Although none of this new fauna possessed hard parts that could be preserved, their soft bodies left impressions in suitable sediments, such as fine-grained mudstones. Various kinds of animal traces are also preserved, such as tracks and burrows. The fact that such tracks are absent in suitable older sediments suggests that relatively complex animals only evolved at this point in the geological record. Many different animal types are represented,

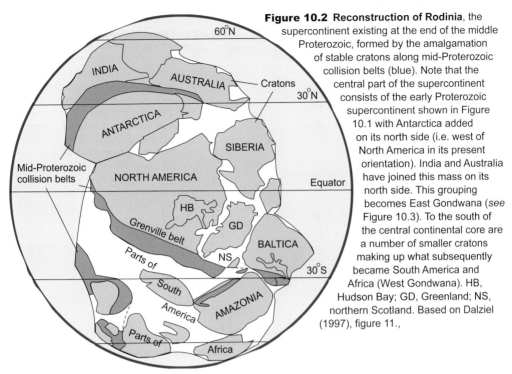

Figure 10.2 Reconstruction of Rodinia, the supercontinent existing at the end of the middle Proterozoic, formed by the amalgamation of stable cratons along mid-Proterozoic collision belts (blue). Note that the central part of the supercontinent consists of the early Proterozoic supercontinent shown in Figure 10.1 with Antarctica added on its north side (i.e. west of North America in its present orientation). India and Australia have joined this mass on its north side. This grouping becomes East Gondwana (*see* Figure 10.3). To the south of the central continental core are a number of smaller cratons making up what subsequently became South America and Africa (West Gondwana). HB, Hudson Bay; GD, Greenland; NS, northern Scotland. Based on Dalziel (1997), figure 11.,

including some that do not appear to have survived into the Palaeozoic Era. The more recognisable forms resemble jellyfish, segmented worms, and **arthropods** ((which include the ancestors of crabs, spiders and insects – *see* Chapter 9). It seems likely that the ancestors of all the main animal groups recognised in the Palaeozoic were represented during latest Proterozoic time.

The supercontinent Rodinia seems to have persisted as a unit throughout the earlier part of this period, and continued subduction of surrounding oceanic crust added the igneous products of volcanic arcs to its margins. However, by about 800 Ma ago, the supercontinent began to split up. The northern part, comprising Australia, Antarctica and India, moved away from North America and collided with the western part of Rodinia, creating a new orogenic belt along the eastern side (in the present orientation) of Africa, termed the **Mozambique belt** (Figure 10.3). The ocean that opened up because of this movement became the ancestor of the modern Pacific Ocean. At about the same time, continental rifts opened up along both margins of Laurentia and between South America and Baltica but did not develop into oceans. One of these became the basin that received the thick sequence of red beds in NW Scotland, known as the **Torridonian**. The record of two widespread glaciations is preserved in the sedimentary basins

10

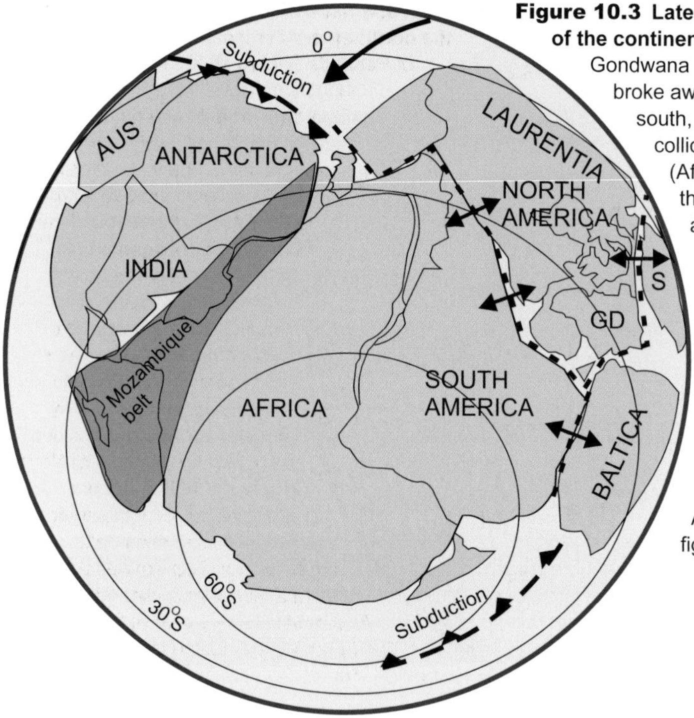

Figure 10.3 Late Proterozoic reconstruction of the continents. About 800 Ma ago, western Gondwana (Australia, Antarctica and India) broke away from Laurentia and moved south, rotating so that its eastern edge collided with eastern Gondwana (Africa and South America) to form the Mozambique orogenic belt and completing the formation of Gondwana. Subduction zones formed along its northern and south-eastern sides. All the continents were now assembled in a single supercontinent (**Pannotia**), but at around 600 Ma ago, Laurentia, Baltica and Siberia broke away. AUS, Australia; GD, Greenland; S, Siberia. Adapted from Dalziel (1997), figure 12.

that formed as a result of these movements, one along the western margin of Laurentia at around 700 Ma, and the other, about 100 Ma later, along the eastern margin, in eastern Greenland, NW Scotland and western Scandinavia.

The new continental configuration, called **Pannotia**, lasted until about 600 Ma ago when the continental mass created by the amalgamation of former north Rodinia (Australia, Antarctica and India) and south Rodinia (Africa and South America) split away from Laurentia to form the new supercontinent of **Gondwanaland** or, more correctly, **Gondwana** (Figure 10.3). The ocean that opened up as a result of this movement is called the **Iapetus Ocean**.

The new supercontinent of Gondwana was to last until the Mesozoic Era (*see* below) when it finally split up to form the present continents of the southern hemisphere. The new orogenic belts created around the margins of Africa and South America as Gondwana was formed are collectively known as the **Pan-African**.

The Lower Palaeozoic (542–416 Ma)

The record of this period commences with the **Cambrian** and **Ordovician** systems. The strata of this age in the lands around the North Atlantic are shallow-water quartz sandstones and marine limestones, now well exposed in eastern North America, NW Scotland and

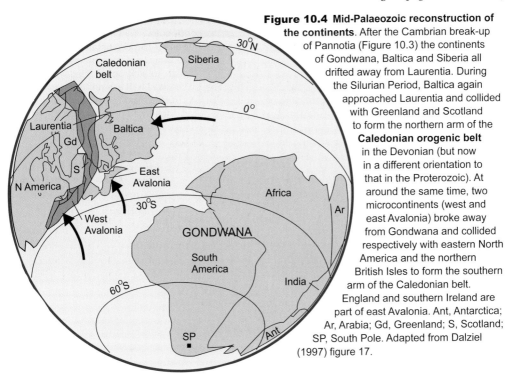

Figure 10.4 Mid-Palaeozoic reconstruction of the continents. After the Cambrian break-up of Pannotia (Figure 10.3) the continents of Gondwana, Baltica and Siberia all drifted away from Laurentia. During the Silurian Period, Baltica again approached Laurentia and collided with Greenland and Scotland to form the northern arm of the **Caledonian orogenic belt** in the Devonian (but now in a different orientation to that in the Proterozoic). At around the same time, two microcontinents (west and east Avalonia) broke away from Gondwana and collided respectively with eastern North America and the northern British Isles to form the southern arm of the Caledonian belt. England and southern Ireland are part of east Avalonia. Ant, Antarctica; Ar, Arabia; Gd, Greenland; S, Scotland; SP, South Pole. Adapted from Dalziel (1997) figure 17.

Scandinavia. Their presence shows that the break-up of Pannotia led to the continental shelves of Laurentia being flooded by shallow seas in which flourished a great diversity of animal life, now leaving calcareous hard parts as evidence. **Trilobites** and **brachiopods** (*see* Chapter 9) are particularly important. Around the margins of the new Gondwana supercontinent, however, quite different conditions prevailed. Here the process of subduction led to the development of deeper-water sediments such as greywacke, together with voluminous vulcanicity. The contrast between these two types of environment can be demonstrated by comparing the Cambro-Ordovician record in NW Scotland (then part of Laurentia) and that of England and Wales (then part of Gondwana): the former dominated by limestones, the latter by greywackes, mudstones and volcanics. The faunas of these two realms also differ, since they belong to different **faunal provinces** (*see* Chapter 9) and were separated by the Iapetus Ocean.

Palaeomagnetic reconstructions show that most of the continental masses lay close to the equator during the Cambrian Period but that, towards the end of the Ordovician, Baltica moved towards Laurentia, whereas Gondwana moved towards the south pole. Glacial conditions now prevailed over much of Gondwana and were accompanied by a **mass extinction** of many marine species (*see* Chapter 9). However,

10

part of northern Gondwana broke away and also moved northwards towards Laurentia. A piece of this terrain, called **Avalonia**, included the southern half of the British Isles.

During the **Silurian** Period, the Iapetus Ocean gradually closed as Baltica approached Laurentia. In the British Isles, the faunas on each side of the closing ocean became more similar towards the end of the Ordovician and into the Silurian. Thick successions of grey-wacke and volcanics both in southern Scotland and south of the closing Iapetus Ocean in northern England, Wales and Ireland indicate that subduction zones now lined both sides of that ocean. The closure of the Iapetus Ocean and the resulting collision of the continental masses of Laurentia, Baltica and the smaller microplate of Avalonia (containing part of western Europe) resulted in the **Caledonian orogeny** (Figure 10.4).

The Caledonian orogenic belt extends from northern Norway southwards through the British Isles and is matched on the western side of the North Atlantic by a similar belt in eastern Greenland, Newfoundland and the northern Appalachians in the USA. By **Devonian** time, the chain of mountains formed by the colli-sion was bordered by large non-marine basins receiving thick sequences of red sandstones and conglomerates known collectively as the **Old Red Sandstone**. West and south of the new Old Red Sandstone continent created by the amalgamation of Laurentia and Baltica, the broad continental shelves were dominated by limestone deposits.

The broad shallow seas of the Silurian Period were rich in animal life, and abundant coral reefs hosted a wide variety of marine fauna. The first jawed fish appeared then, some of which were as large as modern sharks.

During the Devonian, many species of fresh-water fish inhabited the shallow lakes of the Old Red Sandstone basins of Scotland, made famous in the early 19th century by the Scot-tish stonemason, Hugh Miller. At this time also, the first insects had appeared and a group of vertebrate animals (the **amphibians**) had evolved limbs and crawled onto the land. The first land plants, which propagated by spores, began to inhabit swamps in Silurian time but by the late Devonian, forests of seed-bearing plants were forming that could colonise dry land. By late Devonian time, the gap between the Old Red Sandstone continent and Gond-wana had narrowed to the extent that there was little difference in the marine fauna bordering both. By the end of this period however, a mass extinction affected these faunas, probably due to a global cooling event evidenced by glacial deposits in parts of Gondwana (*see* Figure 5.2).

The Upper Palaeozoic (416–251 Ma)

The **Carboniferous** and **Permian** periods saw further major tectonic movements and climatic changes. During the Carboniferous, Gondwana approached and finally collided with the Old Red Sandstone continent, forming orogenic belts where Africa met Europe and North America. The **Hercynian orogenic belt** was formed in Europe, the **Mauretanides** in North Africa and the southern **Appalachian** and **Ouachita** belts in the USA (Figure 10.5). By the end of the Palaeozoic, Siberia had also collided with Europe to form the **Urals orogenic belt** and, with other minor additions, the supercontinent of **Pangaea** had been com-pleted (*see* Figure 5.2).

The continental shelves in the early Car-boniferous (the **Mississippian** to American

10

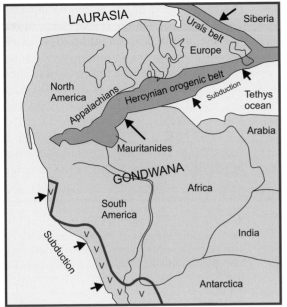

Figure 10.5 The assembly of Pangaea. During the late Carboniferous Period, the supercontinent of Gondwana again approached the combined continents of Laurentia and Baltica (Europe). Northern Africa collided first with southern Europe to form the **Hercynian orogenic belt**, extending from the southern British Isles to the Caspian Sea, then moved west to collide with eastern North America to form the southern **Appalachian belt** in the USA and the **Mauretanide belt** in North Africa. Further east, Siberia collided with Europe along the **Urals orogenic belt**. A subduction zone was formed along the southern margin of the combined continents of Europe and Siberia at the northern margin of the Tethys ocean, and subduction also occurred along the western margin of Gondwana to form a volcanic arc (V–V). Adapted from Matte (1986) figure 17A.

geologists) were dominated by limestones. However, the coral reefs of the Devonian had largely disappeared – **crinoids**, **brachiopods** and **molluscs** (*see* Chapter 9) were now important. The latter half of the Carboniferous (the American **Pennsylvanian**) was marked by a general lowering of sea level, probably related to the widespread glaciation of the southern hemisphere. Broad swamps, which existed in places in the early Carboniferous, became abundant in the late Carboniferous, ultimately forming the bulk of the world's coal deposits. The coal-forming swamps were populated by primitive trees up to 30 metres in height, underlain by ferns and mosses. By now the first flying insects had evolved and the colonisation of the land by vegetation allowed more advanced vertebrates, the **reptiles**, eventually to become abundant in the Permian. At this time, the combined continents of North America and Europe were situated near the equator, whereas the Gondwana continents were clustered around the south pole and glaciers covered much of their landmass. The coal-forming swamps were not confined to low latitudes however, as they also formed in the northern parts of Gondwana and in Siberia, which lay nearer the north pole (*see* Figure 5.2).

During the **Permian** Period, the climate seems to have become drier, and although Gondwana was now attached to Laurasia, the strong contrast between the cold polar and hot equatorial climates caused marked differences in the floras to arise between the northern and southern parts of the supercontinent. These differences were to become more marked as the continents diverged during the Mesozoic. The arid conditions in the tropical regions led to the formation of widespread evaporite deposits – notably the great salt deposits of

10

the central USA and Northern Europe, including England. Continental sand dune deposits are also common. The late Permian also heralded a mass extinction of species, possibly the greatest there has ever been. This may be linked to the fact that, as the supercontinent extended from the north to the south pole, both ends experienced glacial conditions, which were accompanied by a marked fall in sea level.

The Mesozoic Era (251–65 Ma)

By the beginning of the Mesozoic, the last major separate landmass, China, had joined the other continents to form the supercontinent of Pangaea. During the **Triassic** Period, therefore, Pangaea occupied a swathe of land extending from the north to the south poles and formed a crescent shape, enclosing a part of the world-wide ocean known as the **Tethys Ocean** (*see* Figure 5.2). Much of this landmass was arid, and desert conditions prevailed like those of the preceding Permian Period. Evaporites were common in the western USA and on the other side of Pangaea, bordering the Tethys Ocean, in North Africa and Europe. These continental deposits gave way to marine sediments along the borders of Tethys in southern Europe.

This large landmass was occupied by a great variety of vertebrates, including the **dinosaurs**, which underwent an evolutionary explosion after the mass extinction at the end of the Permian. Small **mammals** also appeared towards the end of the Trias. Separate floral provinces characterised the southern, central and northern parts of Pangaea – caused by the equatorial position of North America and Europe, while southern Gondwana and Siberia experienced polar conditions. The

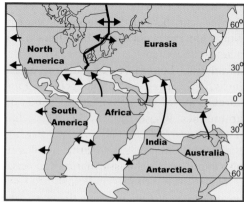

Figure 10.6 The break-up of Pangaea. Pangaea started to break up during the Jurassic, when the central Atlantic was formed. This mid-Cretaceous reconstruction shows that, by then, Africa had separated from South America; by the Miocene, it would collide with Europe to form the **Alpine orogenic belt** (*see* Figure 5.6). The combined continents of India, Antarctica and Australia have separated from Africa. By the late Cretaceous, India would break away from Antarctica and move north towards Asia, colliding with that continent and forming the **Himalayan orogenic belt** by Miocene times. Australia would separate from Antarctica during the Eocene and collide with Indonesia also in the Miocene. North America moved westward away from Europe during the Eocene. During the Mesozoic, subduction zones existed along the western coasts of the Americas and along the southern and eastern margins of Eurasia. Based on Smith & Briden (1977), map 7.

forests of this period were populated by ferns and a group of primitive trees (including the conifers) which carried exposed seeds.

By the end of the Triassic, Pangaea began to break up, with the spreading westwards of the Tethys seaway, eventually to separate North America from Africa in the **Jurassic** Period (Figure 10.6). Sea level rose during the Jurassic, and shallow seas spread inland from the Tethys Ocean into central and western Europe, and flooded large areas of western North America. Widespread limestones characterise these continental shelves, hosting a variety of bottom-dwelling marine fossils, including **bivalves**, **gastropods**, **echinoids** and **corals**. Free-swimming predators include the **ammonites**, whose rapid evolution provides a useful means of dating the various stages of the Jurassic. Bony fish and marine reptiles are also found in these Jurassic marine deposits.

While these continental shelf deposits were forming elsewhere, along the west coast of North America, subduction of the ancestral Pacific Ocean was taking place, resulting in a chain of volcanoes and the addition of numerous island arcs to the continent. Because the direction of subduction was oblique to the continental margin, these new pieces of crust, after docking with the continent, were carried northwards along major strike-slip faults. Many of these so-called **terranes** ended up in Alaska, much of which is composed of these displaced terranes. The evidence for the exotic nature of such terranes comes from their more southern faunas and the magnetic signatures in their rocks.

During the **Cretaceous** Period, the process of fragmentation of Pangaea continued. The central Atlantic widened, the South Atlantic opened up, separating South America from Africa, and India, Antarctica and Australia moved away from Africa (Figure 10.6). The separation of these continents was accompanied by voluminous outpourings of lavas. By the end of the Cretaceous, Europe became detached from Laurentia with the opening of the North Atlantic, and of the Gondwana continents only Antarctica and Australia were still joined together. For the first time in

geological history, large areas of oceanic crust have been preserved; oceanic crust of most earlier periods has been destroyed by subduction. Because this oceanic crust can be dated, the movements of the continents during the Cretaceous Period, and subsequently, can be reconstructed much more accurately than was possible for previous periods.

Sea levels during the Cretaceous were higher than today's and the continents were extensively blanketed by shallow marine deposits. In the upper part of the Cretaceous, much of these deposits was in the form of chalk, including the famous **Chalk** beds of southern England. Chalk is a type of pure white limestone largely formed from the skeletons of tiny unicellular **planktonic** algae (**coccoliths**) and **foraminifera** (*see* Chapter 9). Other fossils commonly found in these deposits include various burrowing bivalve molluscs, gastropods and oysters. Shallow-water reefs were built by **rudist molluscs**. Deep-ocean muds contain siliceous **diatoms**. Cretaceous seas also hosted various arthropods such as crabs and giant marine reptiles, some up to ten metres in length.

Land vertebrates included the spectacular dinosaurs, whose fossil remains have been recorded in large numbers in the late Cretaceous continental deposits of the central USA, Canada and China; these include both herbivores and carnivores. Flying reptiles and birds are also represented. The recent discovery of feathered dinosaurs in China provides an important link between dinosaurs and birds.

The end of the Cretaceous Period is marked by dramatic changes in both tectonic conditions and climate and also, spectacularly, by the most famous (although not the most severe) mass extinction of much of Mesozoic organic

10

life. Although associated popularly with the demise of the dinosaurs, this extinction event also affected many other groups of plant and animal life, including ammonites and marine vertebrates. Most species of planktonic life were extinguished and, since they were at the bottom of the food chain, this had a knock-on effect on the higher organisms. It is believed by many geologists that the extinction is related to a collision with a large meteorite (or even a comet) and that dust from this impact spread around the globe causing disastrous climatic changes. Others point to the widespread increase in vulcanicity around the same time, which could also have had a major environmental impact. The occurrence of a layer of dust containing grains that show evidence of high-velocity impact favours the meteorite collision theory.

The Cenozoic Era (65–0 Ma)

The **Cenozoic Era** is divided into three periods (**Palaeogene, Neogene** and **Quaternary** – *see* Table 8.1) and seven **epochs**:

- **Palaeocene** (65–56.8 Ma);
- **Eocene** (56.8–33.9 Ma);
- **Oligocene** (33.9–23.0 Ma);
- **Miocene** (23.0–5.3 Ma);
- **Pliocene** (5.3–1.51 Ma);
- **Pleistocene** (1.51–0.01 Ma);
- and **Holocene** (or **Recent**) (100,000 years–?).

The **Palaeocene** and **Eocene** Epochs were marked by several major tectonic movements (Figure 10.6). Australia broke away from Antarctica and moved north, ultimately to collide with Indonesia on the southern margin of Asia; India also moved north towards Asia, and Africa moved towards Europe, partially closing the Mediterranean Sea. In North America,

continued subduction of Pacific ocean crust beneath the western continental margin resulted in widespread vulcanicity and the development of the **Cordilleran orogenic belt**, responsible for the spectacular Rocky mountains of the western USA and Canada. The volcanic belt resulting from this subduction zone has continued to the present day in the form of the Cascades Range of northwestern USA and western Canada, but further south, from around 38 Ma ago, the plate boundary has been marked by the **San Andreas transform fault** (*see* Figure 5.7).

Between 55 and 34 Ma ago, during the Eocene Epoch, the Indian continent finally made contact with Asia, beginning a series of events that were to culminate in the main Himalayan orogeny at 20 Ma in the Miocene. The consequences for the tectonics of Asia were profound: over a width of several thousand kilometres, central Asia is crossed by several great mountain chains in addition to the Himalayas, such as the Pamirs, Tien Shan and Altai ranges; it also hosts the high plateau of Tibet and great inland basins such as the Tarim – all the result of this collision.

In Europe, the main **Alpine orogeny**, resulting from the collision of Africa with Europe, took place towards the end of the **Oligocene**, at around 28 Ma, causing the development of a series of mountain chains around the northern margin of the Mediterranean Sea. Further east, Arabia converged with Asia to form the Zagros mountains of Iran. Clastic sediments of Oligocene and Miocene age derived from these mountains formed in continental basins landward of the mountains; these non-marine clastic deposits are termed **molasse.** At the same time, thick marine clastic deposits (termed **flysch**) were laid down offshore.

10

By the **Miocene** Epoch, from about 20 Ma onwards, the present-day tectonic framework of the world was established (*see* Figures 5.6, 5.7). Vulcanicity was concentrated along the ocean ridge network (e.g. in Iceland), around the margins of the Pacific Ocean (the 'Pacific ring of fire') and along the northeastern boundary of the Indian Ocean. In Africa, the great **African Rift** valley system together with the **Red Sea–Gulf of Aden Rift** (Figure 5.8) had become established, with its associated vulcanicity, linked to a hotspot beneath East Africa.

The first two epochs of the Cenozoic, the Palaeocene and the Eocene, were times of relatively high sea level and warm climates. After the mass extinction at the end of the Mesozoic, considerable expansion of marine and terrestrial life took place. Marine life now resembled that of the present day and, on land, although the terrestrial vertebrate faunas were relatively primitive, all the main present-day families were represented. The warm climate extended well into polar regions and fossil palm trees are known from Alaska. However, equatorial regions were not much different – there appears to have been almost no climatic gradient from the equator to the polar regions.

A major climatic change in the late Eocene seems to have been prompted by the separation of Australia from Antarctica which, by isolating Antarctica – then centred around the south pole – caused a change in the oceanic current system, diverting warm waters away from Antarctica. This is believed to have caused the start of the glaciation of the south polar region, which in turn affected the whole world by causing a drop in sea level. In the Arctic too, a change in the ocean circulation pattern was caused by the separation of Asia and North America, which previously had been joined by a land bridge across the Bering Straits. During the Oligocene therefore, global cooling and the lowering of sea level meant that most of the continents were above sea level and grasslands replaced forests over much of North America, with consequential effects on the faunas.

Pliocene to Recent (the last 5 Ma)

The Miocene epoch ended 5.3 Ma ago and was succeeded by the **Pliocene**, during which the climate became warmer and sea level rose, causing flooding of continental margins in western and eastern North America and along the Mediterranean coasts. This warming only lasted for 2.5 Ma, and was ended by the start of the **Pleistocene** glaciation. This event, usually known as the '**Ice age**' was actually a long series of glacial episodes separated by warm periods (**interglacials**). The latest of the glacial episodes ended between 15,000 and 10,000 years ago, and it is assumed by many geologists that we are currently experiencing one of the warm interglacial periods. In fact, past interglacial periods have been warmer than at present. Each change from glacial to interglacial episode would have been accompanied by considerable changes of flora and fauna, as organic life retreated ahead of the spreading glaciers and advanced again when climatic conditions improved. Many of the large vertebrates of the Pleistocene, familiar because of their preservation in peat deposits or frozen soil, are now extinct – such as the **woolly mammoth**, the **mastodon** and the **sabre-tooth tiger** – but most are still with us today. The rise of Man dates from the beginning of the Pliocene, when members of the

10

Hominid family (the genus *Australopithecus*) first appeared. The genus *Homo* dates from the early Pleistocene whereas modern man, *Homo sapiens,* may only be around 100,000 years old.

At its furthest extent, the Pleistocene glaciation covered a large part of the globe. Continuous ice stretched from the northern USA through Canada and Greenland across the north Atlantic to the British Isles and Scandinavia, and thence across much of central Asia. Smaller ice caps existed in the southern hemisphere, centred on southern South America, southernmost Africa, SE Australia and New Zealand, in addition to an expanded Antarctic ice cover. This volume of ice naturally resulted in a lowering of sea level, which became depressed by about 100 metres below the present level. This exposed the continental shelves to erosion, which cut deep valleys through the soft sediment to produce what now are termed **submarine canyons.** Inland, deep valleys were created which were subsequently flooded when the ice retreated, giving rise to the typical fjord-type coastline seen in western Scotland and Norway.

As the ice retreated at the end of the last glacial episode, the typical glaciated landscapes appeared, as seen for example in northern Britain and Canada. The effects of glaciation are described in Chapter 3, and include ice-scoured rock surfaces, U-shaped glacial valleys, ice-dumped **moraine** and glacial clay deposited in ice-dammed lakes

(*see* Figures 3.7, 3.8). The weight of the great ice sheets resulted in the depression of the land surface, which is still in the process of recovering. Thus Scotland and Scandinavia are rising at a rate of several millimetres per year compared to southern England and the southern borderlands of the North Sea, which are slowly sinking. The post-glacial rise of these northern lands is shown by the widespread occurrence of **raised beaches** around the present coastline (*see* Figure 3.5A). In Scotland, old shore lines can be traced at 5, 8, and 16 metres above sea level, and even higher examples appear in Scandinavia.

The period since the end of the last glacial episode (i.e. about the last 10,000 years) is referred to as the **Recent** (or **Holocene**) **Epoch**. Since we do not know whether there are to be more glacial episodes or not, we should perhaps leave it to future observers, if such still exist, to decide the nomenclature of post-Pliocene time.

On a final note, there has been much discussion and speculation about the existence and extent of man-induced global warming. Whatever the extent of this phenomenon, it will be hard, from a geological standpoint, to measure against the extremes of climatic variation throughout geological time, even through the Pleistocene. But another worrying phenomenon, which is easy to measure and predict, is the mass extinction of species which Earth is currently experiencing as a result of human exploitation.

10

Geology and industry

Most people will be aware of the connection between geology and the oil and coal industries, and probably that geological expertise is required to find and exploit metallic ores such as copper, uranium, etc. What may not be so obvious is that whole fields of geology are devoted to civil engineering and water resources. The industrial applications of geological science constitute **economic geology**. This chapter is therefore divided into sections dealing respectively with: **oil and natural gas**; the **extractive industry** (coal, building materials, etc.); **ore geology**; **civil engineering** (the building industry); and **hydrogeology** (groundwater resources).

Oil and natural gas

Natural **petroleum** occurs in the form of liquid oil, gas or solid (**asphalt**). It consists of a mixture of hydrocarbon compounds plus variable amounts of sulphur, nitrogen and oxygen. Petroleum is formed as a result of the partial decomposition of planktonic marine organisms (algae) that have been deposited on the sea floor under reducing conditions (cut off from oxygen). There the algae are decomposed by bacteria to form an organic mud which, when buried and subjected to heat and pressure, produces petroleum. The oil and gas is then squeezed out of the **source rock** by the pressure of the overlying rock and migrates upwards to concentrate in a suitable **reservoir rock**. Oil reservoir rocks require to be both **porous** and **permeable**; that is, they must have open pore spaces to accommodate the oil and gas and these spaces must connect together via channel ways. To contain the petroleum, the reservoirs must also be sealed by a **cap rock** that is impermeable, such as shale or evaporite (e.g. salt or gypsum). Suitable geological structures that commonly form seals, known as **traps**, include **anticlines**, **fault structures**, **unconformities** and **salt domes** (Figure 11.1).

Most of the known oil reserves are held in anticlinal traps in reservoir rocks of Mesozoic or Cenozoic age, and at relatively shallow depths (less than 2.5 km). Large **oil fields** are located in the Persian Gulf region, Central Asia, the southern USA, Alaska, Venezuela and West Africa. Significant deposits also occur around the British Isles, in the North Sea and on the Atlantic continental shelf. The latter deposits

11

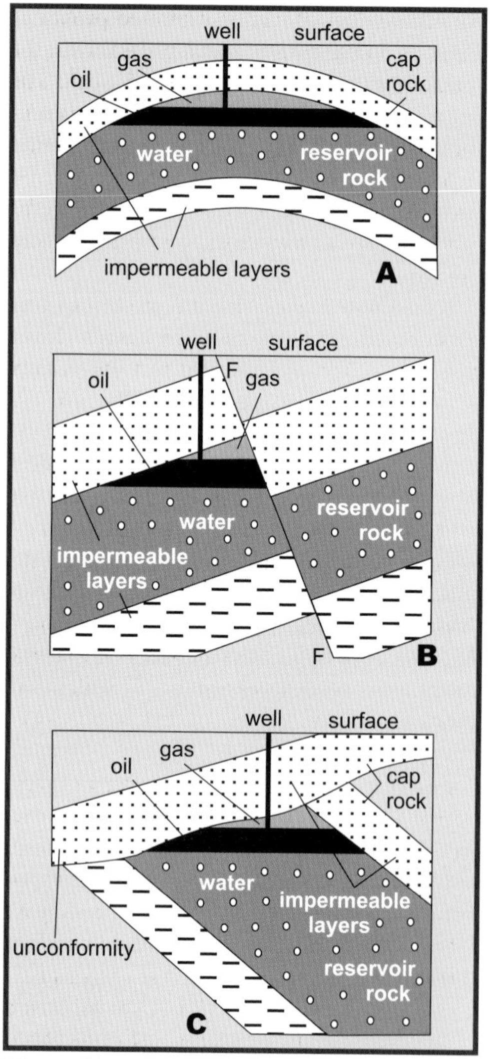

Figure 11.1 Structural traps for oil and gas.
Oil, gas and water are held in a porous, permeable reservoir rock in which the oil forms a layer above the water, with the gas above the oil at the top of the reservoir, held in by an impermeable cap rock. The shape of the impermeable cap is determined by the local geological structure, e.g. an anticline (A), a fault F–F (B), or an unconformity (C).

are derived from either Carboniferous or Jurassic source rocks and held in Mesozoic sandstone reservoirs.

The first petroleum deposits were found by observing oil seeping out at the surface. Since that time, efforts to discover and develop oil fields have become ever more sophisticated. As the price of oil has increased, and as many of the older fields are nearing exhaustion, it has become economically more viable to search for and exploit much smaller fields such as those in the North Sea.

Subsurface oil traps are detected by various indirect methods, the most widely used employing artificially generated **seismic waves** (*see* Chapter 6). This technique is termed **seismic surveying** and is particularly useful at sea, where an array of receivers (**geophones**) is towed behind a ship and the waves generated by a small explosion are reflected back to the geophones from the various rock layers and structures beneath the surface. Once a potential trap is found, test drilling is carried out by a **drilling rig** to determine the presence and extent of the petroleum reserves in the trap. Pieces of broken rock produced by the drill bit are carried up to the surface by the returning lubrication mud and examined in order to determine the local stratigraphy. This process is termed **mud logging**. Alternatively, solid **drill core** can be brought up for examination, or various indirect methods can be employed to measure the physical properties of the rock using instruments lowered down the borehole.

Recovery of the oil or gas takes place either naturally, under its own pressure, or by pumping it out. Injection of water or gas may also be undertaken to recover more of the reserves. Producing oil wells vary from

Figure 11.2 An oil production platform at Invergordon dock, Ross-shire, Scotland. Note the size of the structure compared to the dockside buildings.

very simple pump engines to huge production platforms of the type situated in the North Sea and the Gulf of Mexico (Figure 11.2).

Ore Geology

Ore is rock material mined or extracted in order to recover useful metals. The ore is typically associated with various amounts of worthless rock, termed **gangue** material, which hosts the more valuable ore. Most metal ores are in the form of sulphides, oxides or carbonates: common examples include **cassiterite** (tin oxide), **chromite** (magnesium-iron-chromium oxide), **chalcopyrite** (copper-iron sulphide), **galena** (lead sulphide), **malachite** (copper carbonate), **pitchblende** (uranium oxide), and **sphalerite** (zinc sulphide). Gold and silver are both found as uncombined metals, known as **native gold** and **native silver**. Whether or

not an ore deposit can be exploited profitably depends on the current metal price and on the concentration of the ore in the deposit. The more valuable metals such as gold are profitable even at very low concentrations (around 10 parts per million) whereas iron, which is relatively abundant and cheap, can only be profitably exploited at concentrations of over about 25% iron.

Most metallic elements are present in very low concentrations in igneous magmas. Only a few, such as iron, magnesium and aluminium, occur in significant proportions. The average basalt contains about 10% iron oxide, 8% magnesium oxide and 15% aluminium oxide, but the more valuable metals, such as nickel, copper, lead and zinc are present only in minute amounts (quite variable but typically in the range 10-100 parts per million). Metals therefore can only be economically exploited when they are concentrated by various natural processes, the more important of which will now be described.

● *Magmatic processes.* The cooling and crystallisation of magma results in **magmatic differentiation** (*see* Chapter 2) by the process of **fractional crystallisation,** where certain early-formed crystals sink to the base of a magma chamber and form layers enriched in certain elements. Heavy metal ores such as magnetite, chromite and ilmenite (titanium oxide) are concentrated in this way. The Bushveld gabbro intrusion of South Africa hosts valuable chromite deposits formed by magmatic differentiation. Separation of sulphide-rich liquids is another source of concentration, responsible for the well-known Sudbury copper–nickel deposits of northern Ontario, Canada. Fractional crystallisation of magmas also leads

11

to concentrations of minor elements in liquids that are left over when the main mass of magma has crystallised. These residues are particularly associated with granites and form veins and irregular rock masses, termed **pegmatites**. These pegmatites are a source of metallic ores (e.g. tin and uranium oxides) whose molecular structure does not fit easily into those of the common silicate minerals, and are hosted typically by quartz and feldspar.

● *Hydrothermal deposits.* These deposits are formed by hot water, usually carrying dissolved salt (**brine**) together with various metallic compounds in solution. The water may be derived from siliceous igneous intrusions (e.g. granite) during the late stages of crystallisation of the magma. Other sources are groundwater, sea water and water driven off from **metamorphic** rocks. These **hydrothermal** solutions concentrate metallic elements during the late stages of igneous crystallisation, and they are also able to dissolve metals from already solid igneous rocks or from the surrounding host rock that they pass through. The hydrothermal solutions percolate through the rock via channels and fissures, crystallising in cooler areas as **mineral veins**. The ore is hosted typically by quartz in siliceous country rocks, or by calcite in limestone areas. The acidic solutions react strongly with limestones in particular, commonly forming sulphides of copper, lead and zinc. Gold and silver ores are also formed in this way, as are many other metallic ores. The heat supply for these solutions is mainly derived from igneous bodies, although some hydrothermal deposits have no known igneous source and may have resulted from warm water squeezed out of deeply buried shales, or from the heat generated during regional metamorphism. An important class of massive sulphide ore deposits originate from the hydrothermal **black smokers** found along the volcanically active zones of ocean ridges (*see* Chapter 5).

● *Sedimentary deposits.* Metallic ores can be concentrated by sedimentary processes as a result of their greater density and resistance to erosion compared with the surrounding sedimentary mineral grains such as quartz, calcite, feldspar and mica. Such deposits are known as **placer deposits** and are a common source of gold and other heavy metals such as uranium and tin. Gold is often recovered from modern rivers by amateur prospectors using the technique known as **gold panning**. Small amounts of gold collect in pockets among deposits of river gravel or sand. This sediment is shaken in a pan to allow the small gold grains to settle in the bottom of the pan. Ancient placer deposits, such as the Precambrian conglomerates of the Witwatersrand in South Africa are an important source of gold, although ancient consolidated sedimentary host rocks are much more expensive to separate from the metal and are generally uneconomic.

● *Residual deposits.* These are formed by chemical weathering at the surface, typically in tropical climates with high rainfall. Under these conditions, soils may form in which the soluble components have been dissolved out leaving insoluble residues enriched in iron (**laterite**) or aluminium (**bauxite**). Such soils form above source rocks rich in alumina and/or iron oxides, such as basic volcanics or granite. Bauxites consisting of high proportions of aluminium hydroxide are the main source of aluminium. Laterites developed above ultramafic igneous rocks have produced

11

economic concentrations of nickel, such as those of New Caledonia.

● *Secondary enrichment.* Some economic ore deposits are formed as a result of enrichment above a primary ore deposit where the proportion of ore may be much smaller. The enrichment takes place because of the action of groundwater. For example, pyrite breaks down to yield insoluble iron hydroxide and sulphuric acid, which dissolves metals in the form of soluble sulphates. These are then carried downwards and precipitated as sulphides in a layer immediately below the water table and above the primary ore body. Above this level, within the lower part of the water table, metals may be deposited as oxides or carbonates. The insoluble iron hydroxide remains at the surface, forming a brown residue termed a **gossan**. Gossans are useful signs to the prospector of the presence of potentially valuable ore beneath. Copper ore bodies are often subject to this type of enrichment, forming deposits of malachite (copper carbonate) and chalcopyrite (copper iron sulphide).

Prospecting methods

Traditionally, valuable ores were discovered by prospectors, who relied upon direct observation of surface outcrops, guided by their own experience of likely sites, using features such as the gossans mentioned above, and checking exposures of quartz or calcite veins for signs of mineralisation. During over 2000 years of exploration, however, most of the obvious ore deposits have been discovered and economic geologists have had to devise indirect methods of surveying for further deposits. Both geophysical and geochemical methods are now widely used, both in the discovery of new ore bodies and in the determination of the limits of known bodies.

Magnetic surveying can be carried out by air and is useful for locating magnetic minerals such as magnetite, **ilmenite** (titanium oxide) and hematite. Variations in **electrical conductivity** are also useful in locating metallic ores, and can be measured from the air as well as by more detailed surveying on the ground. Geochemical methods look for unusually high concentrations of metals in soils, river sediments, river water or even vegetation. Ore bodies can be located by following high concentrations in river tributaries upstream until the source area is found. Geochemical surveying can also be carried out by aircraft.

Extraction

Most ores are now recovered by **opencast mining**, whereby the ore is extracted by quarrying in huge pits. The rock is then crushed and the ore extracted by various physical or chemical methods to separate the metals from the worthless gangue material. Underground mining was much used in the past for valuable ores such as gold, copper or lead but is much less employed at present because it is generally uneconomic. A well-known exception is the underground gold mines of South Africa, where the ore is mined to a depth of several kilometres.

The extractive industry

Various kinds of rock are routinely extracted for industrial use: for example, as fuel (coal), building stone, aggregate for road building, lime for cement, and sand and gravel for concrete production.

● *Coal.* This well-known substance is the result of the decay of vegetation in the absence of oxygen, when the material called **peat** is

11

formed. As a result of burial, compression and increased temperature, coal is formed due to the removal of water and other volatiles. About five metres of peat are required to form a one-metre thick layer of coal. The quality of the coal depends on the depth of burial and the temperature of formation and varies from **lignite** and **brown coal**, which contain more volatiles, to **anthracite**, which has the highest proportion of carbon and the lowest of volatiles. Coals are found in rocks of Devonian to Cenozoic age in every continent. The most valuable coalfields are of Upper Carboniferous age in Europe, Asia and North America, and of Permian age in the Gondwana continents. Coalfields of Mesozoic age are also of economic importance in North America and Australia. Coal is extracted both by opencast methods, in large quarries, and by underground mining, although the latter has now become less economic in many countries, including Britain.

● *Building stone.* Rock suitable for building stone is not as common as might be thought. A good building stone is required to be free from undesirable weathering or alteration products, or minerals that would be susceptible to weathering or pollution by sulphur dioxide, for example. It should also be hard and have high mechanical strength. Igneous rocks such as granite have been widely used because of their hardness and resistance to weathering, but suffer the disadvantage of being difficult (and expensive) to work. Limestones are widely used because of their relative ease of working but are susceptible to chemical weathering. Sandstones are also widely used but can be porous, and often contain impurities such as iron compounds, which may weather to produce unsightly brown staining.

Slate (metamorphosed mudstone) was widely used in the past as a roofing material because of its impermeability, but has largely been replaced by artificial materials.

● *Cement.* This material, vital to the construction industry, is made by heating crushed limestone with clay to produce a fine powder. This then recrystallises to a rock-like consistency when water is added. Large amounts of relatively pure limestone are required to produce cement in sufficient quantities, and quarries are a familiar part of the landscape in areas of Carboniferous limestone in the British Isles, such as the Pennines and central Ireland.

● *Sand and gravel.* These materials are used together with cement in the production of concrete. Most economically viable deposits of sand and gravel are unconsolidated, and may be either of **fluvio-glacial** origin (deposited from melting Pleistocene ice sheets), or **alluvial**, deposited in modern river valleys.

● *Aggregate.* This term is used for rock crushed to gravel size for use in road building. Suitable rock needs to be hard, with a high crushing strength. Limestones and fine-grained igneous rocks are widely used for this purpose.

● *Rock salt and gypsum.* Rock salt (sodium chloride) is used in the chemical industry and also for keeping roads free of ice in winter. Gypsum (hydrated calcium sulphate) also has many industrial uses, including the manufacture of plasterboard. Both salt and gypsum (in its anhydrous form, **anhydrite**) are extracted from **evaporite** deposits such as those of Permian age in central England. In many hotter parts of the world, salt is recovered from sea water which is held in shallow tanks and allowed to evaporate.

11

● *Industrial clays.* Clay has many industrial uses including brick and tile making, pottery and ceramics. Mudstones and shales contain variable proportions of **clay minerals**, which are a group of hydrated silicates, with a sheet structure similar to mica (*see* Chapter 1), which become plastic when wet. A good industrial clay should contain a suitable proportion of clay minerals to enable it to be successfully worked and fired: brick clays require a lower proportion of clay minerals than ceramics, for example. The firing process drives off the water and transforms the material to one composed of hard crystalline silicates. Most British industrial clays come from Carboniferous and Mesozoic mudstones. The purest British clay, which has long been used in the ceramics industry, is derived from hydrothermally-altered, deeply weathered Cornish granites. This **china clay (kaolinite)** is extracted from the weathered rock by washing, which separates out the clay minerals from the accompanying quartz and mica.

Diamonds

Diamonds are valued industrially because of their exceptional hardness, optical qualities and also, of course, as gemstones because of their brilliance. The mining of diamonds has many similarities to the mining of valuable metal ores, in that rocks with very low proportions of diamond can be profitably worked because of the very high value of the product. Diamond, however, is not an ore but simply the element **carbon** in its high-pressure crystalline form. It occurs in a type of igneous rock, termed **kimberlite**, which is formed at great depths within the mantle. Kimberlite is an **ultrabasic** igneous rock (*see* Chapter 2) with a higher than usual proportion of potassium-bearing minerals such

as mica. It occurs in small pipe-like bodies, typically only about a kilometre across, in ancient Precambrian rocks, but most of the kimberlites themselves are relatively young – less than about 200 Ma old. The most productive diamond deposits occur in Australia, western and southern Africa, and Siberia.

The diamonds occur in very low concentrations. In the well-known Kimberley mine in South Africa, for example, the proportion of diamond to host rock is only about one part in eight million. Other mines might have higher proportions of diamonds but the proportion of gem-quality diamonds is the more important factor in determining profitability.

Diamonds also occur as **placer deposits** in rivers draining from a kimberlitic source area.

Geology in civil engineering

Any building project requires geological input to a greater or lesser degree at some stage. The building materials themselves require to have the appropriate strength for the purpose, and all major building work should be preceded by a survey to ensure that the ground conditions are suitable for the type and scale of the building in question. The preliminary survey may employ both traditional geological and indirect geophysical methods designed to determine the type of rock, its strength and permeability, and the existence of any fractures or areas of weaker rock that may affect the overall strength. A series of boreholes may be drilled to provide more detailed information. Where there is a cover of superficial deposits, the depth to bedrock is one of the more obvious concerns. The criteria used to determine the suitability of the site depend on the size and nature of the structure: the foundations of a large heavy structure are obviously

11

more critical than those, say, of an ordinary dwelling house.

The strength of the rock material is of critical importance and much effort is devoted to measurement and testing of the various rock types encountered. The most suitable rock types for foundations are crystalline igneous or metamorphic rocks, limestones and most sandstones. Clays and shales are less suitable because they are weaker, and are susceptible to expansion when wet and shrinkage when dry. Faults and heavily fractured or jointed rock must either be avoided, or dealt with by stabilisation with cement (**grouting**) for example.

Dams and tunnels require particularly careful geological preparation, as do road and railway construction where artificially steep slopes are needed, such as embankments and cuttings. The design of dams is vitally important because of the potentially catastrophic results of failure. Dams in narrow, steep-sided valleys are preferred because of the large ratio of water volume to surface area in the reservoir. However, because of the enormous weight of the water behind the dam, the strength of the structure is critical and must depend heavily on the suitability of the ground geology. Any rock type or structure that could allow water to leak through the dam, or to otherwise destabilise it, must be avoided. Since many valleys follow faults, these must always be a potential problem; not only because they are a source of weakness but also because of the danger of earthquakes. Tunnels need protection from rock falls or water ingress where they cross faults or poorly consolidated rock, for example, by lining with concrete. Unstable slopes are often created during road or railway construction and must be stabilised by various means,

such as terracing, or building retaining walls, or enclosing in wire netting. The geological environment of such oversteepened slopes is therefore important. For example, the cutting of slopes across bedding that dips down-slope should be avoided because of the danger of rock masses slipping down bedding planes.

Hydrogeology

Water is an absolutely critical resource. Those parts of the world with high rainfall obtain their water supply from surface water via rivers or reservoirs, but where rainfall is inadequate to supply the needs of the population, other sources of water must be relied upon, such as **groundwater**, or the **desalination** of sea water. The branch of geology that deals with groundwater resources and supply is called **hydrogeology**.

Groundwater is held in porous and permeable rocks such as sandstones and limestones, and also in unconsolidated deposits of sand and gravel. Bodies of rock that contain, or are capable of containing, water resources are termed **aquifers**. The level beneath which water is present in a given area is termed the **water table**. This surface is not necessarily horizontal, but varies with the topography, rising beneath hills and falling in valleys to meet the water level in the occupying river or lake (Figure 11.3). Prospecting for water depends on a knowledge of the local geology – for example where suitable permeable layers or faults might be located. Instruments measuring electrical resistivity can be used to indicate the presence of water beneath the ground, and many farmers rely on water diviners who employ traditional methods using hand-held metal wires or willow branches, which can also lead to successful results.

11

In arid and semi-arid regions, or indeed any regions suffering from a shortage of surface water, knowledge of the geology of the local aquifers is critical. Some aquifer beds extend for great distances and rise into hilly ground where the water table is higher than the area where the water is required (Figure 11.3). In such a case, the water, when tapped by boreholes, may come out under pressure. Such aquifers are termed **artesian**, and wells tapping such a resource are called artesian wells. When the water pressure decreases, water has to be pumped out of the aquifer. However, care must be taken to ensure that too much water is not extracted from the aquifer, since that will affect the water supply further up-dip in the aquifer. In some areas there is the added danger that salt water or other contaminants can affect the water supply if extraction is not carefully controlled.

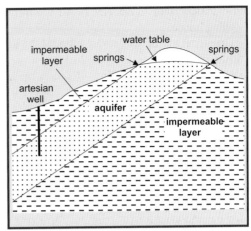

Figure 11.3 An **artesian aquifer** showing how rainfall in hilly ground can be transferred along a permeable rock layer to be tapped under pressure by an artesian well.

11

G

Glossary

A

abrasion [24]: the mechanical rubbing or wearing away of solid rock produced by the action of water or ice carrying rock fragments.

absolute age (dating) [74]: (establishing) the age of a rock stated in 'real time' units (i.e. years before the present) measured **radiometrically**, rather than in **stratigraphic** units.

acid (igneous rock) [16]: an igneous rock containing **quartz** formed from a magma that is oversaturated in **silica**.

African rift (system) [103]: a zone of rift valleys extending from the Red Sea coast in Ethiopia in the north to the Indian Ocean coast in Mozambique in the south, characterised by **vulcanicity** and extensional **fault** movements.

agate [7]: a form of coloured microcrystalline **quartz**, where the colours are typically arranged in bands; found in **geodes**; regarded as a **semi-precious stone**.

aggregate [110]: material composed of angular pieces of broken rock used in the building industry e.g. for road construction.

alabaster [37]: a fine-grained compact form of **gypsum**, used for ornaments.

algae [79]: primitive organisms of two main types – **blue-green algae**, which are single-celled organisms without a nucleus, and which flourished in the **Precambrian**, and the more advanced green, red and brown algae, which include the sea-weeds, and are the simplest forms of plant life; all algae contain **chlorophyll**, consume solar energy, water and carbon dioxide, and produce oxygen as a waste product.

alluvial (deposit) [110]: a sedimentary deposit formed by a river, particularly during flooding; an alluvial plain is the result of an extensive flood or series of floods.

Alpine orogeny [102]: orogenic episode affecting southern Europe, resulting from the collision of Eurasia and Africa during the **Cenozoic**, and culminating in the **Miocene**.

amber [8]: yellow fossil resin derived from **Cenozoic** coniferous trees; often contains fossil insects; regarded as a **semi-precious stone**.

amethyst [6]: purple crystalline form of **quartz**, valued as a **gem stone** (*see* Figure 1.2C).

ammonoids [85]: an extinct group of the **Class Cephalopoda**, possessing coiled chambered shells with partitions joined to the shell by often complex wavy lines (sutures); they include the ammonites (*see* Figure 9.2).

amphibia [88]: a class of **vertebrates** with limbs, which are able to move about on land but require to lay their eggs in water; they include frogs and newts; they are believed to provide an evolutionary link between the **fish** and the **reptiles**.

amphibole [8]: member of a group of ferromagnesian **silicate** minerals with complex compositions; **hornblende** is the commonest variety.

andesite [16]: fine-grained **intermediate igneous** rock (*see* Table 2.1).

anhydrite [110]: the mineral calcium sulphate ($CaSO_4$), found in **evaporite** deposits.

Animalia [79]: the Kingdom Animalia includes all forms of animal life, both **vertebrates** and invertebrates.

anthracite [110]: a type of **coal** characterised by a high proportion of carbon and a low proportion of volatile constituents.

anticline [65]: a **fold** where the folded layers are in the form of an arch with the older beds in the centre (*see* Figure 7.1A).

Appalachian (belt) [98]: **orogenic** belt along the southeastern margin of North America; the northern part, from New York to Newfoundland, resulted from the **Caledonian orogeny**, of **Ordovician** to **Devonian** age; the southern part, reaching as far south as Alabama, resulted from the late **Carboniferous Hercynian orogeny** (*see* Figure 10.5).

aquifer [112]: a rock body containing, or capable of containing, water.

Archaean Eon [74]: unit of geological time 4000-2500 Ma ago (*see* Table 8.1).

artesian [113]: a type of **aquifer**, or well tapping an aquifer, in which the water pressure is sufficient to drive the water to the surface unaided by pumping (*see* Figure 11.3).

Arthropoda [86]: a **phylum** of invertebrate animals characterised by having an external skeleton and jointed limbs; it contains by far the largest number of **species** of any phylum and includes insects, crabs, spiders and centipedes, as well as the extinct **trilobites**.

asphalt [105]: solid **petroleum**, also known as tar or bitumen.

asthenosphere [49]: weak layer beneath the **lithosphere**, capable of solid-state flow, over which the **plates** move.

asymmetric folds [65]: a set of folds with **limbs** of unequal length (*see* Figure 7.1C).

Avalonia [98]: a small continental fragment consisting of parts of eastern North America and western Europe (including England), that was detached from **Gondwana** during the early **Palaeozoic** and collided with **Laurentia** during the **Silurian** (*see* Figure 10.4).

Aves [116]: *see* **birds**.

B

bacteria [82]: primitive, one-celled organisms without a nucleus; bacteria were among the earliest recognisable organisms.

Baltica [94]: a continent consisting of most of northern Europe that existed during the Lower **Palaeozoic**, prior to the **Caledonian orogeny** (*see* Figure 10.4).

banded iron formation [92]: a finely-layered iron-rich sedimentary rock, mostly confined to the **Precambrian**; the commonest type consists of interbanded iron oxide and **chert**.

barite [4]: also known as barytes; mineral composed of barium sulphate (BaSO4); found in **hydrothermal** veins, sometimes with **ore** minerals or **calcite**.

basalt [16]: fine-grained **basic igneous rock**, composed typically of **feldspar**, **pyroxene** and **olivine**, found mainly as **lava** flows.

basic (igneous rock) [16]: igneous rock without **quartz**, formed from a **magma** that is undersaturated with **silica**; the ferromagnesian minerals are typically **pyroxene** and **olivine**.

batholith [14]: a **pluton** with a very large horizontal extent and great thickness, often with no determinable floor (*see* Figure 2.4B).

bauxite [108]: a **residual deposit** composed of aluminium hydroxide formed by **chemical weathering** under tropical conditions; bauxite is the principal **ore** of aluminium.

bed [34]: a layer of **sedimentary rock** representing either a single episode of sediment deposition, or a period of intermittent deposition of the same type of sediment, and bounded above and below by surfaces representing interruptions of sedimentation.

bedding plane [34]: the surface separating adjoining **beds**.

G

bedded chert [37]: *see* **chert** (*see also* Figure 4.4).

beryl [7]: crystalline mineral composed of beryllium aluminium silicate; coloured varieties such as **emerald** are valued as **gems**.

biotite [4]: dark coloured, ferro-magnesian variety of **mica**.

birds [88]: members of the **vertebrate Class Aves**; birds are warm-blooded, and possess both wings and feathers; they are believed to be related to the **reptiles**.

Bivalvia [86]: a **class** of the **Phylum Mollusca** consisting of mussels, oysters etc. which possess two opposed shells and are typically fixed to the sea floor.

black smoker [49, 108]: hydrothermal **vent** situated along the central volcanically active zones of **ocean ridges**. Deposits of ore such as metallic sulphides are formed by chimney-like structures around the vent.

blue-green algae [82]: primitive, microscopic single-celled organisms without a nucleus, which flourished particularly in the Precambrian; also known as **cyanobacteria**; *see* **algae**.

boudin [69]: a sausage-like shape produced as the result of the stretching of a layer (*see* Figure 7.4B).

boulder clay [33]: an unstratified glacial deposit consisting of boulders and rock fragments of varying size in a clay matrix; also known as **till**.

boulder conglomerate [36]: a conglomerate composed mainly of boulders greater than about 25 cm across.

brachiopod [84]: solitary marine animals with two opposed shells belonging to the Phylum Brachiopoda.

breccia [36]: a **clastic** sedimentary rock composed mainly of angular fragments greater than 2mm in diameter.

brine [108]: a solution of common salt (sodium chloride) in water; such solutions often contain other salts and are important in the transport and deposition of many minerals, including **ores**.

brown coal [110]: *see* **lignite**.

buckling [65]: a process of **folding** produced by compression acting approximately parallel to the layers.

butte [29]: a steep-sided, pillar-shaped, isolated hill formed by **erosion** in a desert or semi-desert environment (*see* Figure 3.6B).

C

cairngorm [7]: a grey variety of crystalline **quartz**; valued as a **semi-precious stone**.

calcite [2]: a common mineral composed of calcium carbonate ($CaCO_3$); it is usually colourless or white and may form well shaped crystals (*see* Figure 1.2B).

Caledonian (orogeny) [98]: an orogenic episode affecting most of the British Isles together with western Norway and eastern Greenland during **Ordovician** to **Devonian** time; the orogeny culminated in collision between the continents of Laurentia, Baltica and Avalonia (*see* Figure 10.4).

Cambrian (Period) [97]: a unit of geological time in the **Palaeozoic Era**, 542–488 Ma ago (*see* Table 8.1).

canyon [29]: a narrow steep-sided erosive channel caused by intermittent 'flash floods' in a desert environment (*see* Figure 3.6A).

cap rock [105]: an **impermeable** rock layer which seals a deposit of **petroleum** preventing it from rising to the surface (*see* Figure 11.1).

carbon [4]: the element carbon (C) occurs naturally as oxides or carbonates and also as a **native element** in the form of **graphite** and **diamond**.

Carboniferous (Period) [98]: a unit of geological time in the **Palaeozoic Era**, 359–299 Ma ago (*see* Table 8.1).

cassiterite [107]: the mineral tin oxide (SnO_2) the most important **ore** of tin.

catastrophism: the belief, popular in the 19th century, that geological phenomena are the result of a single catastrophic event (such as the biblical Noah's flood), or a relatively small number of such events, rather than

G

the slow, gradual changes envisaged by the **uniformitarian** theory.

cement [38, 110]: a mineral (e.g. **calcite**, **quartz**, **hematite**) deposited in the pore spaces of **sedimentary** rocks; in industry, a material composed of a mixture of clay and calcium oxide used in building.

Cenozoic Era [102]: unit of geological time; subdivision of the **Phanerozoic Eon**, 65 Ma to present (*see* Table 8.1).

Cephalopoda [86]: a class of the **Phylum Mollusca**; the cephalopods are fast-swimming marine predators with large brains, eyes and tentacles, and include the squids and octopuses as well as the extinct **ammonites**.

chalcopyrite [5]: mineral composed of copper-iron sulphide ($CuFeS_2$); an important source of copper.

Chalk [38, 101]: a soft, white, fine-grained **limestone** of upper **Cretaceous** age in northwest Europe; it is composed mostly of the calcite skeletons of unicellular organisms.

chemical sediment [37]: a sedimentary deposit resulting from a chemical process, such as precipitation from solution.

chemical weathering [22]: weathering produced by the chemical action, particularly of acid rainwater, on rock, which causes alteration to the minerals leading to the gradual disintegration of the rock.

chert [37]: a rock consisting of microcrystalline **silica**, in the form of sedimentary layers or as nodules within **limestones**; formed either by the accumulation of microfossils or chemically, by precipitation.

chevron fold [65]: a fold with straight **limbs** and a sharp angular **hinge** (*see* Figure 7.1B).

chilled margin [15]: the fine-grained margin of an **igneous intrusion** caused by the more rapid cooling and crystallisation of the **magma**.

china clay [111]: a pure white plastic clay used in the ceramic industry, formed by the **hydrothermal** alteration of **feldspars** in **granite**.

chlorophyll [128]: green substance present in plant cells enabling **photosynthesis** to take place.

Chordata [79]: a **phylum** of the animal kingdom containing the **vertebrates** – mammals, birds, reptiles, amphibians and fish.

chromite [107]: the mineral iron-chromium oxide ($FeCr_2O_4$), the major **ore** of chromium.

cirque [33]: *see* **corrie**.

class [79]: a group of related organisms consisting of a number of **orders** (*see* Table 9.1).

clast [24]: a fragment of rock produced by **erosion**.

clastic (rock) [36]: resulting from the disintegration of older rocks.

clay mineral [22]: a group of silicate minerals with a varied chemical composition (mostly hydrated aluminium silicates), characterised by a sheeted structure and the property of plasticity when wet.

club mosses [89]: a group of primitive **vascular** plants, important in the **Carboniferous**, that reproduced by means of spores; some reached heights of 30 metres.

coal [37]: a sedimentary rock, used as a fuel, consisting mainly of plant remains in various stages of decay, ranging from **brown coal** (organic- and volatile-rich) to **anthracite** (carbon-rich).

cobble conglomerate [36]: a conglomerate composed mainly of large fragments around 10–15 cm across.

coccoliths [88]: single-celled, microscopic, marine **algae** with a spherical shape, covered in calcareous plates; coccoliths are important in the formation of **limestone** in the **Mesozoic** and **Cenozoic**.

coelenterates [84]: a group of primitive, jelly-like, marine invertebrates consisting of a hollow body cavity surrounded by tentacles for catching prey; they include **corals**, jellyfish and sea anemones.

columnar jointing [64]: type of cooling joint found in **lava** flows where the rock is divided into columns by sets of polygonal joints (*see* Figure 6.8).

G

conglomerate [36]: a **clastic** sedimentary rock composed mainly of rounded fragments larger than 2 mm in diameter.

conservative (plate boundary) [49]: a **plate** boundary characterised by sideways motion against the adjoining plate along a **transform fault**.

constructive (plate boundary) [48]: a **plate** boundary where new plate is created by divergent motion of two adjacent plates, marked by an **ocean ridge** or **continental rift**.

contact metamorphism [52]: thermal metamorphism caused by the heat generated by an **igneous** body.

continent [42]: in a geological context, the landmass of a continent plus adjacent sea bed underlain by continental-type crust, including the **continental shelf** and **continental slope**.

continental drift [42]: the theory of the relative movements of continents around the Earth's surface.

continental rift [49]: a **constructive plate boundary** situated on continental **crust** and marked by an elongate valley, usually volcanically active.

convection current [19, 44]: a pattern of flow in liquid or solid material driven by a temperature difference; this produces a density imbalance which provides the force necessary to generate the flow.

copper [5]: metal element (Cu) found as a **native element** and as sulphides (**chalcopyrite**), carbonates (**malachite**) etc. in **mineral veins**.

coral [84]: the corals are a group of marine animals with calcareous cup-like skeletons belonging to the **Phylum Coelenterata**; both solitary and colonial forms exist; colonial forms make up coral reefs.

Cordilleran (belt) [102]: **orogenic** belt situated in western North America resulting from a series of **subduction**-related events and the collision of microplates during the **Mesozoic** and **Cenozoic Eras**.

core [17]: the innermost part of the Earth, from a depth of about 2,900km to the centre; the core is composed mainly of iron and a small proportion of nickel; the outer part is liquid (*see* Figure 2.6).

corrie [33]: a deep, semi-bowl-shaped, depression carved out of a mountain by the ice at the source of a valley glacier (*see* Figure 3.8B); known also as **cirque** or **cwm**.

corundum [6]: very hard mineral composed of aluminium oxide (Al_2O_3); coloured forms such as **ruby** and **sapphire** are **gems**; found in **pegmatite** bodies.

craton [93]: the stable part of a continental interior, unaffected by contemporary **orogenic** activity.

Cretaceous (Period) [101]: unit of geological time in the **Mesozoic Era**, 145–65 Ma ago (*see* Table 8.1).

crevasse [30]: a deep open crack or fissure in a glacier (*see* Figure 3.7B,C).

crinoid (Class Crinoidea) [85]: an invertebrate group of the **Phylum Echinodermata**, consisting of a body with long arms attached to the sea floor by a long stalk; the disk-shaped calcareous plates of the stalk are a common constituent of many **limestones.**

cross bedding [38]: structure in coarse-grained **clastic sediments** such as **sandstones**, consisting of a set of tilted **beds** cut by a younger set, caused by changes in direction of water currents or wind (*see* Figure 4.3A).

crust [17]: the uppermost layer of the Earth; the oceanic crust consists of **basalt**,the continental crust of a variety of **igneous**, **sedimentary** and **metamorphic** rocks with an average composition similar to **granite**.

cwm: *see* **corrie**.

cyanobacteria [82]: *see* **blue-green algae**.

D

daughter (element) [74]: an element formed as the end product of a radioactive decay process.

debris fan [23]: a fan-shaped deposit of **scree** formed at the mouth of a steep gully at the base of a cliff or steep slope.

G

deep-focus earthquake [57]: earthquake originating at a depth of below 300km.

deformation [65]: the process whereby rocks are physically altered by the effects of forces acting on or within the **crust**; such effects include **folds**, **faults** and **fabric**.

desalination [112]: the process of removing salt from sea water.

destructive plate boundary [48]: a plate boundary characterised by the convergent movement of adjoining plates, and by the destruction of oceanic **crust** or the collision of continents.

Devonian (Period) [98]: a unit of geological time in the **Palaeozoic Era**, 416–359 Ma ago (*see* Table 8.1).

diamond [6]: high-pressure crystalline form of the element carbon (C); valued as a **gem stone** for its extreme hardness and brilliance; found in **kimberlite** bodies.

diatoms [89]: a group of microscopic, unicellular, marine **algae** with a siliceous external skeleton that form deep-sea siliceous deposits.

dinosaurs [88]: an extinct sub-group of the **Class Reptilia** that were abundant during the **Mesozoic**; they include both carnivorous and herbivorous forms, some were the largest animals that ever existed.

diorite [16]: coarse-grained **intermediate igneous rock** (*see* Table 2.1).

dip-slip fault [60]: a fault where movement has taken place up or down the fault surface (*see* Figure 6.3).

dolerite [16]: medium-grained **basic igneous rock**, composed typically of **feldspar**, **pyroxene** and **olivine;** found mainly as **dykes** and **sills** (*see* Table 2.1).

dolomite [37]: a mineral composed of calcium-magnesium carbonate; also a **sedimentary** rock composed mainly of the mineral dolomite.

dormant (volcano) [12]: currently inactive.

drill core [106]: cylindrical sections of rock obtained by drilling.

drilling rig [106]: a tower-like structure housing a rock drilling machine and drill pipe; it may be located on land or rest on the sea bed on long tubular legs.

drop stones [33]:pebble-size moraine debris deposited on the sea floor by floating ice.

drumlin [33]: a small hill consisting of glacial deposits, moulded by the action of ice passing over it into a smooth rounded shape with its long axis parallel to the direction of ice travel.

dune bedding [38]: large-scale cross bedding in desert sands.

dyke [13]: a body of **igneous rock,** with a sheet-like form and typically steep attitude, that cuts across the structure of the **host rock**, formed by **magma** filling a fissure (*see* Figures 2.3A, 2.4A).

dyke swarm [13]: a set of **dykes** originating from a **magma** source; the dykes may be arranged radially close to the source but become parallel at a distance from it.

E

earthquake [55]: a set of vibrations experienced at the Earth's surface resulting from fault displacement or volcanic activity at depth.

earthquake intensity [55]: the severity of an earthquake as experienced at the surface, measured usually on the **Mercalli scale** (*see* Table 6.1).

earthquake magnitude [55]: the severity of an earthquake, measured as the amount of energy released, as calculated from the size of the resulting **earthquake waves**, using the **Richter scale** (*see* Table 6.1).

earthquake wave [55]: the set of vibrations travelling through the Earth, which are released at the source of an **earthquake**; these are of three main types: **primary (P-) waves, secondary (S-) waves** and **surface waves** (*see* Figure 6.2A).

Echinodermata [84]: a **phylum** consisting of marine invertebrates, including starfish, sea urchins and **crinoids**, characterised by five-sided symmetry and an external skeleton of calcareous plates beneath an outer skin; some

G

possess spines and tube-like feet; they have a relatively advanced nervous system.

Echinoidea [85]: a **class** of the **Phylum Echinodermata**, consisting of the sea urchins, which are mobile grazing animals with moveable spines, living on the sea floor (*see* Figure 9.4B).

electrical conductivity [109]: the degree to which a material conducts an electric current; this property is used as a method of prospecting or surveying; by passing a current through the ground, materials of contrasting conductivity can be distinguished or mapped.

emerald [7]: green crystalline **gem** variety of the mineral **beryl**.

Eocene (Epoch) [102]: a subdivision of the **Palaeogene Period**, 55.8–33.9 Ma ago.

Eon [91]: primary unit of geological time (e.g. the **Phanerozoic**), subdivided into **eras**.

epicentre [55]: the point on the Earth's surface directly above the origin (**focus**) of an **earthquake**.

Era [91]: a unit of geological time (e.g. the **Palaeozoic**), consisting of a number of **periods** (*see* Table 8.1).

erosion [23]: the process whereby particles or fragments of rock are removed from the main rock mass by the actions of rain, wind and ice.

esker [31]: a long narrow ridge of sediment, mainly sand and gravel, deposited by a sub-glacial river.

evaporite [4, 37]: a sedimentary deposit formed by the evaporation of water containing salts (especially sodium chloride) in solution. Other constituents include magnesium chloride, and calcium and magnesium sulphates.

F

fabric [68]: a set of new structures or a texture produced in a rock as a result of **deformation** – e.g. **foliation** or **lineation**.

family [79]: a group of related organisms consisting of a number of **genera** (*see* Table 9.1).

fault [59]: a rock fracture across which appreciable movement has taken place.

fault breccia [64]: a fault rock consisting mainly of breccia (>30% of coarse angular rock fragments).

fault gouge [64]: a fault rock with <30% visible fragmental material in a clay or silt matrix.

faunal province [89]: a geographical area characterised by a particular assemblage of fossil species, and which differs from that of other geographical areas, during a given interval of geological time; such provinces are typically separated by physical barriers such as oceans or mountain ranges.

feldspar [2]: the most abundant mineral constituent of igneous rocks; it is an alumino-silicate with a range of chemical compositions depending on the presence and amount of sodium, potassium and calcium. Orthoclase is the potassium-rich variety; plagioclase ranges from sodium-rich to calcium-rich.

fish [86]: cold-blooded **vertebrates** living in water and breathing through gills; they form several **classes** including primitive jawless types, early armoured fish, sharks and rays, and modern bony fish.

flint [38]: a type of **chert** commonly used by Stone Age peoples for making arrow heads and axes etc

flood plain [25]: a plain formed by successive deposition of sediment from a river in times of flood.

fluorite (fluospar) [4]: mineral composed of calcium fluoride (CaF_2); occurs in hydrothermal veins with lead and zinc ores, **barite**, **calcite** or **quartz** (*see* Figure 1.3B).

fluvio-glacial [110]: a type of process or deposit arising from the activity of melt-water streams or rivers on, beneath, or in front of, a glacier.

flysch [102]: marine **clastic** deposits (typically including **turbidites**) derived from an active mountain range or island arc.

focus [57]: (of an **earthquake**) the location of the origin.

fold [65]: a geological structure formed by the bending of a layer (e.g. a **bed**) as a result of **deformation**.

G

foliation [69]: a set of planar structures produced in rock as a result of **deformation** (*see* Figure 7.4A).

foraminifera [82]: a group of **protozoans** with calcareous shells pierced by tiny holes from which protrude thread-like feelers used for locomotion and feeding.

fore-arc [50]: the part of an oceanic **subduction zone** lying between the **ocean trench** and an **island arc**.

foreland [61]: that part of the continental **crust** lying immediately adjacent to an **orogenic belt** and which has not been significantly affected by it.

formation [34, 73]: a set of **beds** that possess **sedimentary** or **palaeontological** features which distinguish it from adjoining formations and which is given a formal **stratigraphic** name.

fossil [77]: the remains or traces of an organism preserved in rock.

fossil succession [78]: the principle, established by early geologists such as William Smith, that there exists a **stratigraphic** succession in which the different stratigraphic levels can be recognised by the fossils they contain.

fractional crystallisation [19]: a process within cooling **magma** whereby the mineral constituent with the highest melting temperature crystallises first, followed by fractions with successively lower melting temperatures.

fractional melting [19]: the process whereby those mineral components of a heated solid rock with the lowest melting temperatures melt first, followed by those with successively higher melting temperatures.

frost shattering [21]: mechanical weathering of rock caused by alternate freezing and thawing which causes the rock to disintegrate.

fungi [82]: simple unicellular or multicellular organisms including mushrooms and moulds, similar to plants but lacking chlorophyll.

G

Ga (giga-annus) [74]: time unit of a thousand million years.

gabbro [16]: coarse-grained **basic igneous rock**, composed typically of **feldspar**, **pyroxene** and **olivine** (*see* Table 2.1).

galena [5]: mineral composed of lead sulphide (PbS); forms shiny grey cubic crystals; commonly occurs in mineral veins along with **calcite** and **sphalerite** (*see* Figure 1.4D).

gangue [107]: worthless rock from which **ore** is extracted.

garnet [7]: member of a group of hard, iron, magnesium or calcium silicate minerals, considered as **semi-precious stones**, found particularly in **metamorphic** rocks (*see* Figure 1.5B).

Gastropoda [86]: a **class** of the **Phylum Mollusca**, consisting of the snails, which are both marine and land-based animals, typically crawling forms with coiled calcareous shells.

gem (stone) [5]: mineral substance, typically crystalline, that is valued for ornamental purposes.

generic name [79]: the first name of a particular organism, followed by the **specific name** – e.g. *Homo sapiens*.

genus (plural genera) [79]: a group of closely related organisms consisting of a number of **species**; the second lowest in the hierarchy of relatedness (*see* Table 9.1).

geode [2]: a hollow cavity within solid rock, lined with crystals (*see* Figure 1.2C).

geophone [106]: an instrument for detecting artificially generated **seismic waves**.

geyser [11]: a jet of hot water, steam and other gases emitted from a hot spring due to volcanic activity (*see* Figure 2.2B).

glacial erratic [31]: a boulder which has been transported by ice and left lying on the ground surface (*see* Figure 3.8A).

glacial striation [33]: a set of grooves scratched on the rock surface by the movement of a glacier; these are used to indicate the former direction of flow.

G

gneiss [52]: a coarse-grained **metamorphic** rock characterised by the segregation of light-coloured minerals such as **quartz** and **feldspar** into bands or lenses separated by dark minerals such as **micas** or **hornblende**.

gneissosity [69]: the **foliation** characteristic of a **gneiss**.

gold [5]: precious metal element (Au) found as a **native element** in **mineral veins** and in **placer deposits**.

gold panning [108]: the technique used by prospectors (often amateur) whereby the heavier gold is separated out by agitation of the gold-bearing sediment in a pan.

Gondwana (formerly **Gondwanaland**) [42]: a supercontinent that existed during **Palaeozoic** time, consisting of the continents South America, Africa, India, Antarctica and Australia (*see* Figures 5.1, 10.4))

gossan [109]: a deposit of **limonite**-rich material formed as a result of the oxidation of sulphide **ore** deposits; gossans are of no economic value but are useful as an indication of more valuable ore beneath.

Gowganda formation [93]: an early **Proterozoic** sedimentary sequence in southern Canada containing glacial deposits.

graded bedding [39]: structure in coarse unsorted **clastic sediment** where the larger clasts have settled to the base of the bed and the smaller to the top; found particularly in **turbidites**.

granite [15]: a coarse-grained **acid igneous rock** containing **quartz**, **feldspar** and **mica** or **hornblende;** found mainly as **plutons** (*see* Table 2.1).

granite–greenstone terrain [92]: a type of early **Precambrian** terrain characterised by a predominance of **granite** interspersed with **greenstone belts**.

granite tor [22]: an isolated upstanding exposure of bare rock surrounded by free faces on all sides and with a rounded shape, caused by the **weathering** and **erosion** of **granite** (*see* Figure 3.1).

graphite [4]: mineral composed of carbon (C), in the form of black shiny flakes; occurs in veins and in some **metamorphic** rocks.

graptolite [82]: member of the Class Graptolithina, an extinct group of simple colonial marine organisms that flourished during the **Palaeozoic** (*see* Figure 9.1).

greenstone belt [92]: a linear to irregularly-shaped assemblage of volcanic and sedimentary rocks situated within granitic terrain of early **Precambrian** age.

Grenville (belt) [94]: a mid-**Proterozoic orogenic belt** situated along the south-eastern margin of North America.

greywacke [36]: an unsorted sandstone composed of a variety of minerals; typically formed by deposition from a **turbidity current**.

grit(stone) [36]: a coarse-grained sandstone; or one made up largely of angular grains.

groundwater [112]: water present within unconsolidated or **permeable** rock beneath the **water table**.

grouting [112]: construction technique of stabilising unconsolidated or fractured rock by pouring in liquid cement.

Gulf of Aden rift [103]: an arm of the Indian ocean caused by the separation of Arabia and Africa and linked with the **Red Sea rift**; part of a **constructive plate boundary**.

gypsum [4]: mineral composed of hydrated calcium sulphate (*see* Figure 1.3A); the fine-grained compact variety is known as **alabaster**.

H

half-life [75]: the period of time in years taken for a radioactive decay process to reduce the amount of the **parent element** to half.

halite [4]: the mineral sodium chloride (**rock salt** – NaCl), usually containing some magnesium chloride in addition; it is an important constituent of sea water which may be extracted from it by evaporation, forming **evaporite** deposits.

hanging valley [33]: a side valley that has been cut off by a glacier cutting down to a lower

G

level in the main valley (*see* Figure 3.8B).

head of water [23]: the vertical distance between the source of a river and its outlet.

hematite (haematite) [5]: mineral composed of iron oxide (Fe_2O_3); hematite ore is rust-red in colour and is the most abundant source of iron (*see* Figure 1.4A).

Hemichordata [82]: a **phylum** consisting of animals with a primitive spine-like structure, including the **graptolites**.

Hercynian (orogeny) [98]: orogenic episode of mid- to late-**Carboniferous** time resulting from the collision between the **Gondwana supercontinent** and **Laurentia**; it affected much of western, central and southern Europe, including southern Ireland and parts of southwest England (the Hercynian belt); in North America, it formed the southern **Appalachian** and **Ouachita** belts and the **Mauretanides** in North Africa (*see* Figure 10.5).

hinge (of fold) [65]: **line or zone along which the limbs of a fold meet.**

Holocene (Epoch) [104]: also known as the **Recent**, is a subdivision of the **Quaternary Period** and includes the present day.

Hominidae [79]: a family of the Order **Primates** (apes, monkeys etc.), containing the genus *Homo* (man).

hornblende [4]: an important rock-forming mineral; a member of the **amphibole** group; it is a complex hydrated alumino-silicate of iron, magnesium, calcium, sodium and potassium, with a range of chemical compositions; it is a common constituent of **intermediate igneous rocks** such as **diorite**.

host rock [13]: the rock into which **magma** is intruded to form an **igneous body**.

hotspot [53]: a part of the Earth's **crust** exhibiting unusually high heat flow and **vulcanicity**, either within a **plate** (as in Hawaii) or on a **constructive boundary** (as in Iceland – *see* Figure 5.11A).

Hudsonian [93]: an early **Proterozoic orogeny** affecting large parts of North America.

hydration [22]: the process of combining water with a substance (e.g. a mineral).

hydrogeology [105]: the branch of geology dealing with the study of water resources.

hydrothermal [108]: relating to hot water; used to describe solutions passing through rocks and also rocks formed by such solutions, such as **mineral veins** and **ore** bodies.

I

Iapetus Ocean [96]: ocean opened during the late **Proterozoic** and closed during the **Caledonian orogeny**.

Ice age [103]: term often used for the most recent (**Pleistocene**) glaciation, but can also refer to any glacial period in Earth history.

ichthyosaurs [88]: an extinct group of **Mesozoic** carnivorous swimming **reptiles**.

igneous [14]: (of rock or process) produced by, or relating to melted rock (**magma** or **lava**).

ilmenite [109]: an iron-titanium **ore** mineral ($FeTiO_3$) found in **basic igneous** rocks.

impermeable [105]: not **permeable**.

index fossil [73]: a fossil that is characteristic of a particular geological time unit and is used to date it.

interglacial [103]: a warm period within a glaciation when the area covered by ice shrinks.

intermediate (igneous rock) [16]: an igneous rock intermediate in composition between **acid** and **basic**, without **quartz** and typically with **hornblende** or **biotite** as the ferromagnesian minerals (*see* Table 2.1).

intermediate-focus earthquake [57]: earthquake originating at a depth of between 60 and 300 km.

intraplate [57]: relating to the interior of a **tectonic plate** (rather than at a plate boundary); thus, e.g. '**intraplate earthquake**'.

ironstone [37]: a sedimentary rock consisting mainly of iron compounds, typically iron oxides (**hematite**, **magnetite**) or hydroxides (**limonite**).

island arc [50]: a partly submerged arcuate mountain range, typically volcanic, situated

G

Glossary

alongside a **subduction** zone at a **destructive plate boundary** (*see* Figure 5.4).

J

jade [8]: a **semi-precious stone** consisting either of the green sodic **pyroxene** (jadeite) or a green **amphibole**.

jet [8]: a hard black variety of **lignite**, derived from fossil wood; used for ornaments.

joint [59]: a rock fracture across which there has been no significant movement.

Jurassic (Period) [100]: unit of geological time in the **Mesozoic Era**, 200–145 Ma ago (*see* Table 8.1).

K

kame [31]: a mound of **moraine** (fluvio-glacial rock debris).

kaolinite [111]: a common type of **clay mineral** composed of hydrated aluminium silicate.

kimberlite [111]: a type of **mica-peridotite** found mainly in **pipes**, and considered to have originated in the deep **mantle**; it is the main source of **diamonds**.

L

laccolith [14]: a **pluton** with an approximately lens-shaped form (*see* Figure 2.6A).

lamination [34]: small scale **bedding**, where individual layers are only millimetres thick.

landform [21]: a natural structure or shape of the Earth's surface produced by geological processes, such as **erosion**, **vulcanicity** or **deformation**.

Lapland-Kola (belt) [93]: an early **Proterozoic orogenic belt** in northern Scandinavia.

lateral moraine [31]: a long ridge of **moraine** formed at the side of a glacier by rock debris eroded from the valley walls.

laterite [108]: a **residual deposit** of iron and aluminium hydroxides formed under tropical weathering conditions.

Laurasia [42]: a supercontinent that existed during **Upper Palaeozoic** time consisting of the greater parts of the continents North

America, Europe and Asia (*see* Figure 10.5).

Laurentia [94]: a **continent** consisting of most of North America, Greenland and NW Scotland that existed during the Lower Palaeozoic prior to the **Caledonian orogeny** (*see* Figure 10.4).

lava [9]: molten rock (**magma**) which is poured out on the Earth's surface from a volcano or fissure.

Laxfordian [93]: an early **Proterozoic orogeny** affecting NW Scotland.

lignite [110]: a brown to black deposit formed from decomposed plant remains; also known as **brown coal**, this material will become **coal** after further burial and **lithification**.

limbs (of fold) [65]: **the parts of a fold on either side of the hinge.**

limestone [37]: a **sedimentary** rock composed mainly of calcium carbonate, which may be derived either chemically by precipitation or from fossils.

limonite: the hydrated form of iron oxide, typically yellow or brown in colour, formed as an alteration product of iron **ore**.

lineation [69]: a set of linear structures produced in rock as a result of **deformation** (*see* Figure 7.4C).

listric fault [62]: a curved **normal fault** whose inclination decreases downwards (*see* Figure 6.6).

lithification: the process of forming solid rock from sediment.

lithosphere [49]: the strong upper layer of the Earth, with an average thickness of about 100 km, including the **crust** and part of the **mantle**; it consists of a number of **plates** that move over the weaker **asthenosphere** beneath.

load cast [39]: structure formed at the base of a **bed** of coarse **clastic sediment** caused by it pressing down into the underlying mud (*see* Figure 4.3C).

longshore drift [26]: the process whereby currents flowing parallel to the shore carry sand to form elongated deposits (sand bars).

Lower Palaeozoic [96]: informal name for the

early part of the **Palaeozoic Era** (see Table 8.1).

M

Ma (mega-annus) [74]: time unit of a million years.

magma [9]: molten rock, which solidifies to form **igneous** rock.

magma chamber [13]: a large space within the crust into which magma is injected and subsequently may undergo **magma differentiation**; the igneous body resulting from the filling of a magma chamber is termed a **pluton**.

magma differentiation [19]: a process caused by partial crystallisation in a cooling **magma**, which causes differences in composition in the resulting **igneous rocks**.

magnetic surveying [109]: a method of surveying or prospecting using instruments to measure variations in the magnetic intensity of rocks or soils; certain metallic **ore** bodies and **basic igneous** rocks, for example, will show marked differences from the surrounding rocks.

magnetite [5]: black magnetic mineral composed of iron oxide (a combination of FeO and Fe_2O_3); it is a minor constituent of many **igneous** and **metamorphic** rocks and is also found concentrated **in mineral veins** and **magma differentiates**.

malachite [5]: mineral composed of hydrated copper carbonate; has a very distinctive bright green colour (*see* Figure 1.4C); occurs as a product of weathering of other copper **ores**.

Mammalia [88]: the mammals, a **class** of the **Phylum Chordata**; mammals are warm-blooded **vertebrates** with a covering of hair, that feed their young with milk from special glands; primitive types lay eggs, the more advanced give birth to live young.

mantle [17]: that part of the Earth's interior between the **crust** and the **core**, composed mainly of rock with an **ultrabasic** composition (*see* Figure 2.6).

mantle drag [53]: the force exerted on the base of a **tectonic plate** by the frictional drag of the underlying mantle (*see* Figure 5.9).

mantle plume [53]: a column of rising hot mantle material inferred to explain **hotspots** (*see* Figure 5.10B).

marble [52]: a **metamorphic** rock consisting mainly of **calcite**, formed from the metamorphism of **limestone**; **dolomitic marble** is formed from **dolomite**.

marsupial [88]: a type of **mammal** whose young are reared in a pouch attached to the mother's body.

mass extinction [90]: an episode of geologically short duration in which large numbers of species become extinct; examples include those at the end of the **Devonian**, the **Permian** and the **Cretaceous**.

Mastodon [103]: a large, extinct, elephant-like mammal that inhabited North America during the **Pleistocene**.

Mauretanides [98]: **orogenic** belt in northwest Africa; part of the **Hercynian** orogenic system (*see* Figure 10.5).

meander [25]: a curved loop in the course of a river, typically one of a series where a river crosses a flood plain (*see* Figure 3.3B).

mechanical weathering [21]: weathering produced by the action of physical processes such as the disintegration due to the freeze-thaw action of ice or the sand-blasting effect of wind.

Mercalli scale [55]: a subjective method of quantifying the intensity of an **earthquake** as experienced at a particular location at the surface on a scale of increasing severity I-XII (*see* Table 6.1).

mesa [29]: a flat-topped, steep-sided hill formed by erosion in desert or semi-desert country (*see* Figure 3.6B).

Mesozoic Era [74]: unit of geological time; subdivision of the **Phanerozoic Eon**, 251–65 Ma ago (*see* Table 8.1).

metamorphic rock: a rock that has undergone **metamorphism**.

metamorphism [51]: the process of transforming a rock by means of a change in

G

temperature and/or pressure; the changes to the rock typically involve recrystallisation either of the same minerals or to a different mineral assemblage that is stable under the new temperature-pressure conditions.

mica [4]: an important rock-forming mineral with a sheet-like crystal structure forming thin shiny flakes; it is a hydrated alumino-silicate of potassium (**muscovite**) which is colourless, or iron-magnesium (**biotite**) which is brown; there are other less common varieties (*see* Figure 1.2D).

millet-seed (grain) [28]: a near-spherical sand grain formed by wind-induced rolling action in a desert.

mineral vein [2, 108]: a body of rock, typically of sheet-like form but may be irregular, consisting of minerals deposited by aqueous solutions; **ore** minerals may be hosted in veins consisting largely of **quartz** or **calcite**.

Miocene (Epoch) [103]: a subdivision of the **Neogene Period**, 23.0–5.3 Ma ago.

Mississippian [98]: term used in North America for the earlier part of the **Carboniferous**.

Moine thrust zone [60]: major zone of thrusts in NW Scotland defining the western boundary of the **Caledonian orogenic belt** (*see* Figure 6.4).

molasse [102]: continental **clastic** deposits (often red) derived from an active mountain range.

Mollusca [85]: an invertebrate **phylum** of animals (both marine and land-based) with an external calcareous shell or shells; they include mobile swimming forms such as the **cephalopods**, crawling forms like the **gastropods**, and fixed **bivalves** such as mussels.

moraine [31]: unsorted rock debris transported by a glacier (*see* Figure 3.7A).

mould [77]: a type of fossil consisting of the imprint of an organism left in rock after the organism itself has been removed or destroyed.

Mozambique belt [95]: a late **Proterozoic orogenic belt** situated in eastern Africa and western Antarctica formed from the collision between west and east **Gondwana** (*see* Figure 10.3).

mud cracks [28]: shrinkage cracks formed in dried-out mud (*see* Figures 3.5C, 4.3B).

mud logging [106]: the process of monitoring and recording information about the rock being penetrated during drilling operations by examining rock fragments brought up in the drilling mud.

mudstone [37]: a **clastic** sedimentary rock composed of microscopic particles less than 0.004 mm in diameter.

muscovite [4]: pale-coloured potassic variety of **mica**.

mylonite [64]: a fine-grained **fault** rock formed at depth under **metamorphic** conditions and typically showing a regular fine banding; associated with major **thrust** zones.

N

Nagssugtoqidian [93]: an early **Proterozoic orogeny** affecting parts of southern Greenland.

native element [4]: element found in an uncombined form, such as **sulphur**, **copper**, **gold** or **silver**.

Neogene (Period) [102]: unit of geological time, in the **Cenozoic Era**, 23.0–2.5 Ma ago (*see* Table 8.1).

normal fault [60]: a **dip-slip fault** whose upper side has moved down the fault surface (*see* Figure 6.3A).

O

oblique-slip fault [60]: a fault where movement has taken place obliquely on the fault plane.

ocean ridge [12]: long, submerged oceanic mountain range, the site of a **constructive plate** boundary (*see* Figure 5.4).

ocean trench [45]: deep marine trough, site of a **subduction** zone and a **destructive plate** boundary (*see* Figure 5.4).

oceanic crust [17]: crust composed mainly of **basalt** formed initially by volcanic activity at **ocean ridges** (*see* Figure 5.5).

oil field [105]: a substantial deposit of crude

G

oil that is being exploited or considered for exploitation.

Old Red Sandstone [98]: red continental deposits, mainly of **Devonian** age, but locally including late **Silurian** and early **Carboniferous** material, formed as a result of the erosion of the **Caledonian** and northern **Appalachian** mountain chains.

Oligocene (Epoch) [102]: a subdivision of the **Palaeogene Period**, 33.9–23.0 Ma ago.

olivine [4]: an important mineral constituent of **basic** and **ultrabasic igneous** rocks; it consists of iron-magnesium silicate with complete substitution between iron and magnesium.

opencast mining [109]: recovery by surface excavation rather than underground.

order [79]: a group of related organisms consisting of a number of **families** (*see* Table 9.1).

Ordovician (Period) [96]: a unit of geological time in the **Palaeozoic Era**, 488–444 Ma ago (*see* Table 8.1).

ore (deposit) [107]: rock containing minerals of economic value.

organic sediment [36]: a sedimentary deposit formed mainly or entirely from organic material (e.g. fossil accumulations) or as a result of organic activity.

orogenic belt [51]: part of the Earth's **crust**, typically a linear zone, that has undergone **orogeny** (mountain building); orogenic belts are formed as a result of **plate** collision which causes crustal thickening, uplift and the formation of mountains.

Ouachita (belt) [98]: orogenic belt of late **Carboniferous** age (part of the **Hercynian orogeny**) in the southern USA; the westward continuation of the southern **Appalachians** (*see* Figure 10.5).

ox-bow lake [25]: a former **meander** that has become cut off from the river.

P

Palaeocene (Epoch) [102]: a subdivision of the **Palaeogene Period**, 65–55.8 Ma ago.

Palaeogene (Period) [102]: unit of geological

time, in the Cenozoic Era, 65–23.0 Ma ago (*see* Table 8.1).

palaeogeography [40]: geographic conditions during an earlier period of geological time.

palaeomagnetism [44]: the study of the magnetic properties of rocks; principally to determine the orientation of their magnetic palaeolatitude and pole position.

palaeontology [77]: the study of fossils.

Palaeozoic Era [74]: unit of geological time from 542 to 251 Ma ago, usually divided into Lower (older) and Upper (younger) parts (*see* Table 8.1).

Pan-African [96]: a late **Proterozoic** to early **Palaeozoic orogeny** affecting large parts of Africa and South America and resulting in the assembly of the **Pannotia supercontinent**.

Pangaea [42]: the supercontinent, consisting of the whole continental landmass, which existed during much of **Upper Palaeozoic** time (*see* Figures 5.2, 10.5).

Pannotia [96]: a late **Proterozoic** supercontinent formed as a result of the detachment of west **Gondwana** from North America and its collision with east Gondwana along the **Mozambique belt** (*see* Figure 10.3).

parallel fold [65]: a fold in which the layers are of approximately equal thickness throughout.

parent (element) [74]: a radioactive element which decays to form a different element (the **daughter**).

peat [109]: a deposit of partly decomposed plant material; peat, when buried and lithified, becomes **coal**.

pegmatite [108]: an exceptionally coarse-grained crystalline rock, of either **igneous** or **metamorphic** origin, typically found in **veins**; pegmatites are usually of broadly **granitic** composition but have formed from volatile-rich fluids; they are an important source of **ore** minerals

peneplain [25]: a near-horizontal surface formed after a long period of **erosion** by a river system.

Pennsylvanian [99]: term used in North America

G

for the later part of the **Carboniferous**.

peridotite [16]: an **ultrabasic rock** containing a high proportion of **pyroxene** and **olivine** (*see* table 2.1).

Period [91]: a unit of geological time (subdivision of an **Era**) (*see* Table 8.1).

permeable [105]: the property of a substance (e.g. a rock) having connected pathways which allow fluids to migrate through it.

Permian (Period) [99]: a unit of geological time in the **Palaeozoic Era**, 299–251 Ma ago (*see* Table 8.1).

petroleum [105]: naturally occurring hydrocarbon compounds, comprising crude oil, natural gas and **asphalt**.

Phanerozoic (Eon) [74, 91]: the unit of geological time from the beginning of the **Cambrian Period** (542 Ma ago) to the present (*see* Table 8.1).

photosynthesis [82]: the process whereby plants convert sunlight, water and carbon dioxide into food, and release oxygen, using the green substance **chlorophyll**.

phylum (plural, **phyla**) [79]: a group of related organisms consisting of a number of **classes** (*see* Table 9.1); the phylum is the highest grouping in the hierarchy of the animal kingdom.

pillow lava [13]: a type of solidified lava characterised by rounded pillow-like or tube-like shapes formed when liquid lava is poured out into or under water; a rapidly cooled skin of lava forms around the still liquid interior of the pillows (*see* Figure 2.3B).

pipe [13]: also known as a **vent**; an **igneous body** with an approximately cylindrical form, the feeder body to a volcano.

pitchblende [107]: a variety of the mineral uraninite (UO_2), an important **ore** of uranium.

placental mammal [88]: type of mammal whose young are nourished within the mother's body by means of a placenta.

placer deposit [108]: an **ore** deposit formed by sedimentary deposition, e.g. in a river bed; heavy minerals such as gold often form such deposits.

plankton [79]: floating organic life, mostly microscopic.

plants (Kingdom Plantae) [88]: plants are multicellular organisms whose cells possess nuclei and which carry **chlorophyll** to achieve **photosynthesis**; most have a **vascular** system to carry fluids throughout the plant.

plate (tectonic) [48]: a relatively stable piece of the **lithosphere** that moves independently of adjoining plates; plate boundaries are of three types, **constructive**, **destructive** and **conservative.**

plate tectonics [47]: the theory that ascribes tectonic processes to the relative movement of the **lithosphere plates**.

Pleistocene (Epoch) [102]: a subdivision of the **Quaternary Period**, 2.5–0.01 Ma ago.

Pliocene (Epoch) [102]: a subdivision of the **Neogene Period**, 5.2–2.5 Ma ago.

pluton [14]: an **igneous** body of large dimensions, both horizontally and vertically.

porous [105]: the property of (a rock) having pore spaces able to accommodate liquid or gaseous material.

pothole [24]: a rounded 'pot-shaped' depression in the bed of a river produced by the **abrasive** action of rock fragments or pebbles swirled around by the flowing current (*see* Figure 3.3A).

Precambrian [74]: rocks or time older than the **Cambrian Period**.

precious stone [7]: a valuable **gem** mineral.

primary (P-) waves [58]: the first set of earthquake waves to arrive at a recording station; they travel through the Earth by a process of alternate expansion and compression of the material through which they are transmitted in the same way as sound waves (*see* Figure 6.2A).

Primates [79]: an order of the **Class Mammalia**, containing the apes, monkeys etc. together with the **Family Hominidae** (man).

Proterozoic Eon [74]: unit of geological time between 2500 and 542 Ma (*see* Table 8.1).

G

protozoa [82]: primitive, single-celled organisms, whose cells contain a nucleus; thought to be the ancestors of all animal life.

pterosaurs [88]: an extinct group of **Mesozoic** flying **reptiles** allied to the **dinosaurs**.

pumice [9]: rock consisting of solidified 'frothy' **lava** containing cavities filled with gas.

pyrite [5]: mineral composed of iron sulphide (FeS_2); forms brassy yellow cubic crystals (*see* Figure 1.4B); found in **mineral veins**.

pyroxene [4]: an important mineral constituent of **basic igneous** rocks, with a wide range of chemical composition; it is basically a silicate of iron, magnesium and calcium, where each of these three elements may vary from 0–100%; there may also be varying proportions of other elements such as manganese.

Q

quartz [1]: an abundant, very hard, mineral composed of **silicon dioxide** (SiO_2); it may form well-shaped hexagonal crystals (*see* Figure 1.2A); it may be colourless, grey (**cairngorm**), purple (**amethyst**) or pink (**rose quartz**); it is a constituent of many sedimentary rocks such as **sandstones**, and of **acid igneous** rocks.

quartzite [1, 38]: a rock composed entirely or mostly of **quartz**.

Quaternary (Period) [73]: unit of geological time in the **Cenozoic Era**, 2.5-0 Ma (*see* Table 8.1).

R

radiolaria [82]: a group of **protozoans** with a siliceous perforated outer covering, shaped like a bell or hat.

radiolarian chert [82]: a siliceous deposit formed from the skeletons of **radiolaria** and found in deep-ocean environments.

radiometric (dating) [74]: the method of dating rocks that uses the rate of decay of particular radioactive minerals by measuring their ratios.

rain prints [28]: a set of depressions left in mud by rain.

raised beach [26]: an old beach above the present shore line formed at a time of higher sea level (*see* Figure 3.5A).

ramp [60]: that part of a **thrust fault** which cuts up through **bedding** (*see* Figure 6.4A).

Recent (Epoch) [104]: an alternative name for the **Holocene**, the current time unit.

Red Sea rift [103]: an elongate depression, occupied by the Red Sea, caused by the separation of Arabia from Africa, and floored by oceanic crust; part of a **constructive plate** boundary.

relative dating [74]: measuring the age of a geological event in relation to another event or events whose date may or may not be known (e.g. with reference to a fossil assemblage).

regional metamorphism [52]: metamorphism generated by regional changes in temperature and/or pressure, caused for example by **orogenic** processes.

rejuvenation [25]: the process whereby vigorous **erosion** is restored by the uplift of the land surface.

relative age (dating) [74]: (establishing) the age of a rock stated with reference to a particular dated event or to a **stratigraphic** time unit where its **absolute age** is not known.

reptiles [88]: a **class** of **vertebrates**, including the modern crocodiles, lizards and snakes, and the extinct **dinosaurs**; reptiles are cold-blooded animals that lay eggs with shells and can live entirely on land.

reservoir rock [105]: a rock which contains a deposit of **petroleum** (oil or natural gas) which has migrated there from its source.

residual deposit [108]: an **ore** deposit formed by **weathering** and near-surface processes causing a concentration of certain elements; **bauxite** and **laterite** are formed in this way.

reverse fault [60]: a **dip-slip fault** whose upper side has moved up the fault surface (*see* Figure 6.3B).

rhyolite [16]: fine-grained **acid igneous** rock, found typically as **lava** flows (*see* Table 2.1).

Richter scale [55]: scale for measuring **earthquake magnitude**, numbered from

G

1 (lowest) to 10 (highest), where each unit represents an increase of x10 in energy released (*see* Table 6.1).

ridge-push force [53]: the force exerted on a **tectonic plate** by the gravitational push exerted by an **ocean ridge** (*see* Figure 5.10).

ripple marks [28]: a set of ridges and troughs formed on the sand or mud surface by the action of water currents (*see* Figure 3.5B).

river system [23]: a system of drainage formed by a river and its tributaries.

river terrace [26]: a flat area alongside a river which has cut down through the deposits left when it was at a higher level.

roche moutonnée [33]: a small rock outcrop that has been moulded by the passage of ice over it such that it has a smooth slope on the upstream side and is rough and steepened on the downstream side due to the plucking action of the ice (from the French 'sheep rock').

rock salt [4]: *see* **halite**.

Rodinia [94]: a late **Proterozoic supercontinent** consisting of most of the existing continental masses, assembled during mid-Proterozoic **orogenies** (*see* Figure 10.2).

rose quartz [7]: a pink variety of crystalline **quartz**; valued as a **semi-precious stone.**

ruby [6]: red crystalline **gem** variety of the mineral **corundum** containing traces of chromium.

rudist [86]: a group of sedentary **bivalves** which formed large reef-building colonies; the lower shell was shaped like a dustbin with the upper forming the lid.

S

sabre-tooth tiger [103]: a large, extinct, carnivorous **mammal** of the cat family that inhabited North America during **Miocene** and **Pliocene** times.

salt dome [105]: a dome-shaped structure formed by the upward migration of salt under gravitational pressure from its source layer.

San Andreas fault (zone) [58, 102]: a major **transform fault** zone situated along the west coast of California, separating the Americas **plate** from the Pacific **plate**.

sandstone [36]: a **clastic sedimentary** rock composed largely of sand-size grains between about 0.5 and 2 mm in diameter.

sapphire [6]: blue crystalline **gem** variety of the mineral **corundum** containing traces of iron and titanium.

schist [52]: a medium- or coarse-grained **metamorphic** rock characterised by the parallel alignment of platy minerals such as **mica**.

schistosity [69]: the **foliation** characteristic of a **schist**.

scree [23]: a deposit of coarse angular rock debris formed along the base of cliffs or steep slopes resulting from the erosion of the rock above (*see* Figure 3.2B).

sea-floor spreading [42]: the theory of the generation of ocean **crust** at **ocean ridges** and its destruction at **ocean trenches** (*see* Figure 5.5).

secondary enrichment [109]: the process whereby an **ore** becomes concentrated by solution and redeposition of the primary ore, e.g. by the circulation of groundwater.

secondary (S-) waves [58]: the second set of earthquake waves to arrive at a recording station; they are transmitted through the Earth by a process of shear, or lateral movement, of the material through which they travel (*see* Figure 6.2A).

sediment(ary rock) [21, 34]: deposit formed either from fragments or particles derived from older rocks; or from chemical or organic processes such as the evaporation of sea water or the accumulation of fossil remains; or from a combination of these.

seismic survey [106]: a technique of exploration using artificially generated **seismic waves**, which react differently in passing through different materials and can be used to map geological structures in three dimensions.

seismic wave: *see* **earthquake wave**.

seismograph [58]: instrument for detecting **earthquake waves**.

semi-precious stone [7]: a **gem stone** that is attractive but not particularly rare or valuable, such as the various coloured forms of **quartz**.

shadow zone [58]: circular or ring-shaped zone of the Earth's surface which seismic waves from a particular earthquake do not reach.

shale [37]: a **laminated mudstone**.

shallow-focus earthquake [57]: earthquake originating at a depth of 0–60 km below the surface.

shear zone [70]: a zone of **deformation** between two rock masses moving in opposite directions; the equivalent of a **fault** at depth (*see* Figure 7.5).

shield volcano [13]: a volcano with an approximately circular or elliptical plan and gently inclined slopes.

silica [1]: silicon dioxide (SiO_2); e.g. *see* **quartz**.

siliceous [79]: composed wholly or mostly of **silica**.

sill [13]: a sheet-like **igneous body** typically with a horizontal or gently inclined attitude formed by the injection of **magma** along a fissure parallel to the prevailing structure (e.g. **bedding**) of the **host rock** (*see* Figure 2.4).

siltstone [37]: a **clastic sedimentary** rock composed mainly of fragments between 0.004 and 0.06 mm in diameter.

Silurian (Period) [98]: a unit of geological time in the **Palaeozoic Era**, 444–416 Ma ago (*see* Table 8.1).

similar fold [65]: a fold where the folded layers vary in thickness but each layer has the same shape (*see* Figure 7.1E).

slab-pull force [53]: the force exerted on a **tectonic plate** by the gravitational pull of a sinking **lithosphere** slab at a **subduction** zone (*see* Figure 5.10).

slate [52]: a fine-grained **metamorphic** rock formed from **shale** or **mudstone** and characterised by **slaty cleavage.**

slaty cleavage [69]: the closely-spaced **foliation** characteristic of a **slate**.

snowball Earth [90]: the hypothesis that holds that, during certain glacial periods in Earth history, ice covered the whole earth rather than just the polar regions.

sorting [24]: the process whereby **clastic sediments** are separated into groups of different clast size by the action of water or wind.

source rock [105]: the rock from which a deposit of **petroleum** originated.

space problem [14]: the problem of determining how space is made to accommodate a large **pluton** within the **crust**.

species [79]: a group of closely related organisms capable (in theory) of interbreeding; the lowest in the hierarchy of relatedness (*see* Table 9.1).

specific name [79]: the unique name given to a species; the name is in the form **generic name** followed by specific name – e.g. *Homo sapiens*.

sphalerite [5]: mineral composed of zinc sulphide (ZnS); forms dark, brownish cubic crystals in **mineral veins**, often with **galena** and **calcite**.

staircase path (of a **thrust fault**) [60]: the shape of a thrust fault that alternately lies parallel to **bedding** and cuts up through bedding on a **ramp** (*see* Figure 6.4A and B).

strata [35]: **beds** of **sedimentary** rock (e.g. *see* Figure 4.1A).

stratigraphy [36]: the branch of geology dealing with rock successions and geological history.

stratigraphic column [73]: the succession of geological time units ordered from oldest at the base to youngest at the top (*see* Table 8.1).

strike-slip fault [60]: a fault where the movement has taken place horizontally along the fault (*see* also **wrench fault** – Figure 6.3).

stromatolite [82]: reef-building structures produced by the precipitation of calcite by primitive unicellular organisms, mainly **algae** and **bacteria**; they are particularly important in the late **Precambrian**.

subduction [50]: the process whereby an oceanic **plate** descends into the **mantle** along a **subduction zone**; part of the subducted plate melts to give rise to a zone of volcanoes on the opposite plate (*see* Figure 5.8A,B).

G

submarine canyon [28, 104]: a channel cut through **sediments** of the continental shelf or slope formed by the erosive action of a **turbidity current**.

sulphur [4]: an element (S) which occurs as a **native element**, e.g. in hot spring deposits, or as metallic sulphides (e.g. **pyrite**) in **mineral veins** or sulphates in **evaporite** deposits.

supercontinent [93]: a large continental mass consisting of several components that formerly, or subsequently, were themselves continents.

superposition (law of) [71]: the principle underpinning **stratigraphy** – that younger **strata** rest upon older.

surface waves [58]: set of **earthquake waves** that travel around the surface of the Earth's crust and arrive at the recording station after the **P-** and **S-waves** (*see* Figure 6.2A).

syenite [16]: a coarse-grained **intermediate igneous** rock containing a high proportion of potassic **feldspar** (*see* Table 2.1)).

symmetric fold [65]: a set of folds with **limbs** of equal length (*see* Figure 7.1C).

syncline [65]: a **fold** where the folded **beds** are in the form of a trough with the younger beds in the centre (*see* Figure 7.1A).

system [74]: the sequence of rocks formed during a particular **period** of geological time.

T

tectonic: (of a process or structure) relating to or caused by major Earth forces e.g. **plate tectonics**.

terminal moraine [31]: a large deposit of **moraine** dumped at the end of a glacier as it melts (*see* Figure 3.7A).

terrane [101]: a piece of crust, with well-defined boundaries, that differs in tectonic history from its neighbours, and may have originated at some distance.

Tethys ocean [42, 100]: the ocean that separated **Laurasia** and **Gondwana** during upper **Palaeozoic** and early **Mesozoic** time (*see* Figures 5.2, 10.5).

thrust fault [60]: a **reverse fault** which is generally gently inclined (<45°) although it may become steepened by subsequent movements (*see* Figure 6.4).

till [33]: *see* **boulder clay**.

tillite [41]: rock consisting of consolidated **till**; an indication of a previous glaciation (*see* Figure 4. 4A).

topaz [7]: crystalline mineral composed of hydrated aluminium silicate with fluorine; found in **granites** and **pegmatites**.

Torridonian [95]: a late **Proterozoic** sedimentary sequence in NW Scotland consisting of red **sandstones** and **conglomerates**.

tourmaline [7]: crystalline mineral with a complex chemical structure, basically an alumino-silicate with boron and fluorine plus varying amounts of other elements such as sodium, magnesium, iron, manganese and lithium; found in **granites** and **pegmatites**.

transform fault [49]: a fault that forms part of a **plate** boundary where the plates on each side move in opposite directions, parallel to the trend of the fault.

trap [105]: in the context of **petroleum** deposits, a structure which hosts such a deposit (*see* Figure 11.1).

Triassic (Period) [100]: a unit of geological time in the **Mesozoic Era**, 251–200 Ma ago (*see* Table 8.1).

trilobites [86]: an extinct marine group of the **Phylum Arthropoda** that flourished in the **Palaeozoic**; they were crawling or swimming animals rather like wood lice, with a central ridge and two side 'lobes'; they possessed a head shield with complex eyes, and an armoured segmented body and tail, each segment carrying a pair of limbs used for walking, swimming, breathing or food gathering (*see* Figure 9.3B).

tsunami [11, 56]: water waves with long wavelengths caused by sudden movements of the sea floor, e.g. caused by **earthquakes**; they travel at high speed and may cause great destruction when they reach land.

G

turbidite [28]: a **sedimentary** deposit formed by a **turbidity current,** characterised by poor **sorting** of the sediment.

turbidity current (or **flow**) [28, 39]: a water current generated by gravity-induced flow, carrying large quantities of **sediment** of varying coarseness in suspension.

U

ultrabasic (or **ultramafic**) [16]: (rock) composed mainly or entirely of the ferromagnesian minerals **pyroxene**, **olivine** or **hornblende**.

unconformity [71]: the structure formed where a younger set of **beds** cut across an eroded older set, indicating a break in the geological record (*see* Figure 8.1).

uniformitarianism (theory): the doctrine, first advanced by James Hutton, that the geological record was the result of a long series of events and processes similar to those of the present day, rather than a brief sudden catastrophe like the biblical flood.

unroofing [64]: the process of stripping off the rock cover above presently exposed rock outcrop; the term is used particularly in the context of unroofing **joints**, which are due to the release of load pressure by this process.

Upper Palaeozoic [98]: informal name for the later part of the **Palaeozoic Era** (*see* Table 8.1).

Urals (belt) [98]: orogenic belt of late **Palaeo-zoic** age situated along the Urals mountains between Europe and Asia, and resulting from the collision between the Siberian continent and **Laurasia** (*see* Figure 10.5).

V

varve [33]: a seasonal layer of sediment deposited in a glacial lake; the layer consists of a coarser part deposited during the summer melting period and is followed by finer mud deposited at other times.

vascular (system) [89]: the method used by plants or animals to transfer nutrient-bearing fluids throughout the organism.

vent [13]: the mouth of a **volcano**; or the **igneous** body feeding a volcano.

vertebrates [86]: a subphylum of the **Phylum Chordata**; they consist of the most advanced types of animal with a backbone and central nervous system, and include the **fish**, **amphibia**, **reptiles**, **birds** and **mammals**.

viscous [9]: sticky, as opposed to runny, as applied to the flow of a liquid.

volcanic ash [9]: a cloud or deposit of fine solid particles ejected explosively from a volcano.

volcanic breccia [9]: a deposit of coarse angular rock fragments ejected from a volcano.

W

wadi [29]: a watercourse in the desert, dry except during occasional 'flash floods'.

water table [112]: the level beneath which water is present beneath the ground surface; it is an uneven surface, varying with topography and rising or falling with rainfall variation (*see* Figure 11.3).

weathering [21]: the process whereby rock is altered and worn away by the physical and chemical action of the weather: e.g. of rain, wind, ice etc.

whole-rock (date) [75]: a **radiometric** date produced by measuring an element ratio in the whole rock rather than in a mineral constituent.

woolly mammoth [103]: a giant extinct member of the elephant family which inhabited the Arctic regions during the late **Cenozoic**.

wrench fault [62]: a fault where the displacement is horizontal and parallel to the trend of the fault (*see* Figure 6.3C).

Z

zircon [7]: crystalline mineral composed of zirconium silicate; found in granites and pegmatites; contains traces of radioactive thorium, uranium and lead, and is used for **radiometric dating**.

G

Further Reading

Introducing Palaeontology – A Guide to Ancient Life (2010) by Patick Wyse Jackson, Dunedin Academic Press.

Basic Palaeontology: Introduction to Palaeobiology and the Fossil Record (2008) by Michael J. Benton & David A.T. Harper, Wiley Blackwell.

Rocks and Minerals (1992) by Chris Pellant, Dorling Kindersley.

Introducing Volcanology – A Guide to Hot Rocks (2011) by Dougal Jerram, Dunedin Academic Press.

Teach Yourself Volcanoes, Earthquakes and Tsunamis (2007) by David Rothery, McGraw Hill.

An Introduction to Geological Structures and Maps (2003) by G.M. Bennison & K.A. Moseley, Arnold.

Plate Tectonics: How it Works (2008) by A. Cox & B.R. Hart, Blackwell.

An Introduction to Economic Geology and its Environmental Impact (1997) by Anthony Evans, Blackwell.

Sedimentary Rocks in the Field: a Colour Guide (2005) by Dorrick Stow, Manson

The Key to Earth History: an Introduction to Stratigraphy (2001) by Peter Doyle, Matthew Bennett & Alistair N. Baxter, Wiley.

An Introduction to Physical Geography and the Environment (2008) by Joseph Holden. Pearson (Prentice Hall).

Useful Websites

www.bgs.ac.uk: British Geological Survey

www.usgs.gov: US Geological Survey

www.ga.gov.au: Geoscience Australia

www.gns.cri.nz: New Zealand Institute of Geological and Nuclear Sciences

www.geolsoc.org.uk: Geological Society of London (mainly for professional geologists and students).

www.geologist.demon.co.uk: Geologists Association (for UK amateur geologists)

www.ougs.org.uk: Open University Geological Society

www.nhm.ac.uk: Natural History Museum, London (good for fossils)